INTERNATIONAL ENERGY AGENCY (IEA)
SMALL SOLAR POWER SYSTEMS PROJECT (SSPS)

THE IEA / SSPS
SOLAR THERMAL POWER PLANTS
– Facts and Figures –

Final Report of the
International Test and Evaluation Team (ITET)

Editors:
P. Kesselring and C. S. Selvage

Volume 4:
Book of Summaries

Dr. sc. nat. Paul Kesselring
Head, Prospective Studies Division,
Swiss Federal Institute for Reactor Research (EIR),
Head, SSPS Test and Operation Advisory Board.

Clifford S. Selvage, BS
Head, SSPS International Test and Evaluation Team.

ISBN 978-3-540-16149-3   ISBN 978-3-642-82684-9 (eBook)
DOI 10.1007/978-3-642-82684-9

CIP-Kurztitelaufnahme der Deutschen Bibliothek
International Energy Agency / Small Solar Power Systems Project:
The IEA, SSPS solar thermal power plants: facts and figures; final report of the Internat. Test and Evaluation
Team (ITET) / Internat. Energy Agency (IEA), Small Solar Power Systems Project (SSPS). Ed.: P. Kesselring
and C. S. Selvage. – Berlin; Heidelberg; New York; Tokyo: Springer
ISBN-13: 978-3-540-16149-3

NE: Kesselring, Paul [Hrsg.]; HST

EDITORS PREFACE

The Project's origin

As a consequence of the so-called "first oil crisis", the interest in solar electricity generation rose sharply after 1973. The solar thermal way of solving the problem was attractive because the main task was simply to replace the fossil fuel by a "solar fuel" in an otherwise conventional thermal power plant -that was at least what many thought at that time. Thus more than half a dozen of solar thermal plant projects were created in the mid-seventies. One of them is the Small Solar Power Systems (SSPS) Project of the International Energy Agency (IEA). It consists of the design, development, construction, operation, test and evaluation of two dissimilar small solar thermal electric power systems each at a nominal power of 500 kW_e.

ITET and TOAB

In order to assist the Operating Agent (DFVLR - Deutsche Forschungs- und Versuchsanstalt für Luft- und Raumfahrt e.V.) in managing the project, the Executive Committee (EC) created two bodies called the "International Test and Evaluation Team" (ITET) and the "Test and Operation Advisory Board" (TOAB). The latter consisted of a group of experts from the different participating countries, meeting three to four times a year to articulate i.a. the technical interests and expectations of the different parties in the project. It was the TOAB that formulated e.g., the test and evaluation program, which for the final project evaluation boiled down to a list of topics -the so-called "deliverables"- that would be covered by the final report.

The ITET was a group of engineers and scientists that did most of the final evaluation work, supported by cooperative organizations such as e.g., DFVLR and contractors. The final result of this major effort is the present series of books, containing facts and figures, discussed and interpreted to fulfill the task defined in the "deliverables".

Information pyramid

In order not to get lost in the vast amount of information provided by the SSPS Project, an efficient information management was thought to be vital. We adopted a solution which we call "information pyramid" and which will be explained in the introduction. Part of the pyramid is the "Book of Summaries", which contains the abstracted contents of all three volumes of the final ITET-Evaluation Report. Thus, it is possible to get a quick overview of the work performed by the ITET members and to find their conclusion.

General Conclusions

What are these conclusions after all?

We certainly know now that the naive picture of the solar-fired, but otherwise conventional, thermal power plant is wrong. Here we have

learned much from our difficulties with the systems as a whole. Our
main technical success has been the good performance of the solar specific components and subsystems, such as e.g., receivers, collectors,
heliostat fields, etc. They fulfilled most of our high expectations.

Thus, generally speaking, although we did not demonstrate routine power
production from a utility's point of view, we were able to contribute
considerably to the technical advancement of the solar thermal technology. Above all we developed confidence in the technical soundness of
the solar thermal approach. Most companies involved in the project
would be ready to go on with a commercially sized plant, provided there
was a customer.

Acknowledgements

The editors, being the heads of ITET and TOAB, would like to thank
their colleagues for all the work accomplished under sometimes necessarily less than ideal conditions. Without the motivation and per-
severance of the ITET-crew and without the positive, critical minds
of the TOAB members, the present series of books would not exist.

The Operating Agent, DFVLR, provided much valuable support throughout the project. It is a pleasure to acknowledge this help as well
as the good services.of the Plant Operating Authority, Cîa. Sevillana
de Electricidad S.A.

The preparation of the manuscript for publication has been another
formidable task, shared by Sandia National Laboratories, Livermore,
and DFVLR again. Here we would like to thank in particular
Miss Melissa McCreery, secretary of the ITET, Mrs. Sallie Fadda from
Sandia and Dr. H. Ellgering of the DFVLR and their collaborators.

Last but not least, our thanks go to the Executive Committee of the
SSPS Project, whose full support -within the limits of a complex
international cooperation between nine countries- is gratefully
acknowledged.

Würenlingen and Livermore, August 1985

P. Kesselring and C.S. Selvage, Editors

BOOK OF SUMMARIES

TABLE OF CONTENTS

5

INTERNATIONAL ENERGY AGENCY / SMALL SOLAR POWER SYSTEMS (SSPS)

EVALUATION REPORTS

1. INTRODUCTION

This introduction to the final evaluation report of the SSPS International Test and Evaluation Team (ITET) is split into two parts: The first part -written by the head of the Test and Operation Advisory Board (TOAB)- gives a picture of the SSPS evaluation effort as seen from the point of view of an observer far away from the project site in a participating country. The second part -written by the head of the International Test and Evaluation Team (ITET)- gives the general project overview.

1.1 The SSPS Project evaluation, as seen by the head of the TOAB

a) Structure and interaction of ITET and TOAB

In retrospect, the most astonishing feature of the SSPS Project to me is that it was possible to integrate the quite different interests of nine countries to the extent that such a large common venture -worth approximately 90 Million DM- could be realized. This general aspect -i.e., the need to integrate different, sometimes conflicting interests- was also of importance when it came to the organization of the project evaluation. It is e.g., reflected in the structure of the ITET. Only its head and the two senior evaluators were direct "employees" of the Project. All other members were seconded by the different countries to the Project. Their selection, in the countries, was only restricted by relatively loose boundary conditions, set by the Project (minimal duration of stay, minimal number of members to be seconded by a country, preference for certain qualification profiles). As a result, the ITET was a frequently changing group of people, differing not only in nationality but also with respect to the background of education and interests. It was held together by the common task.

While ITET members during their stay in Almería worked full time and on site for the project, many members of the Test and Operation Advisory Board (TOAB) devoted a few days per year only to SSPS activities. The members of this board were designated by their countries in order to help the project with their professional expertise and at the same time to articulate the interest of their countries on a technical level. Thus, in making a main contribution to the definition of the evaluation program, the TOAB selected from the very large number of imaginable R&D subjects, a small fraction, lying within reach of the ITET and reflecting the technical priorities as well as the national interests.

The interaction between TOAB and ITET -whose head and senior evaluators as well as the OA, took part in TOAB meetings ex officio- led to very beneficial side effects: The ITET, struggling with the daily on site problems, could not forget about the needs of the far away home countries and the sometimes (too) high expectations of the TOAB were brought down to the reality of the hard facts in Almería.

b) Structure and character of the ITET Final Evaluation Report

Thus, the stage was set for the final evaluation. It was carried out in the following way: The evaluation topics defined by the "deliverables" were discussed within the ITET and subtasks assigned to the members of the group. The responsibles for each subtask then became the authors of a self-consistent paper, describing their work, results and conclusions. It is the collection of all these individual papers -written within the common framework explained before- that forms the main body of the Final Evaluation Report of the ITET.

The history of the report makes it clear that one should not expect a homogeneous document, covering every possible R&D aspect of the two plants in a comprehensive way. What we must expect and find, is consistency between the different contributions and their conclusions. A variation in the depth and quality of treatment is obvious and finds its natural explanation in the fact that the spectrum of authors begins with engineers, recently graduated from engineering schools, and ends with professors from technical universities. The reader, missing a paper on a topic of high interest to him, must be reminded that time and resources were limited and obviously any selection of priorities is debatable to some degree.

c) The Information Pyramid

Even considering these restrictions, the present volumes contain a very large amount of very valuable information concerning solar thermal power plants. In order to manage this information avalanche efficiently, we have introduced a hierarchy of publications, which we call the "information pyramid". It begins at the top with a book giving a synthesis of the SSPS work in the context of solar thermal power plant development in general. The book makes reference to the present collection of papers frequently. It also appears in a Springer edition and is written by an author hired by the project. The language is such that students and young engineers will be able to follow and the mature engineer gets a quick overview of the important aspects of the solar thermal technology.

The reader, willing to go into more detail, may then take the "Book of Summaries", containing the abstracts of all the papers included in the 3 volumes of the ITET Final Report. He thus has the possibility to decide quickly which of the references given in the book are most important to him and whether or not he should dig into the thick volumes in order to study the complete papers. Complete papers make reference to SSPS Technical Reports and/or to the lowest level, the SSPS Internal Reports. This information as well as raw data are available upon request via the Executive Committee members. Thus, there is a simple and efficient way to get down from the most general, highly aggregated information into more detail, step by step, to end up with the raw data, if necessary.

In parallel to the ITET's Report, the Operating Agent's point of view of the SSPS Project is given in his final report (SR-7: SSPS - Results of Test and Operation, 1981 - 1984).

d) Lessons learned

Concluding my part of the introduction, I would like to give a short, personal view of the lessons learned from the existing solar thermal power plants in general and the SSPS Project in particular. Such statements are necessarily simplifying and incomplete but nevertheless, useful in characterizing the status of a development at a given point in time [1]:

-The development of the solar specific components and subsystems of solar thermal power plants during the last 8 years has been a technical success. Receivers, collectors and heliostat fields perform to a large extent as expected.

-The problems arising from solar specific systems aspects have been underestimated. We mention in particular:

·Start-up, shut-down time of plant (transient behavior)
·Heat energy management in storage systems
·Troubles with "from the shelf" components and subsystems.

They have been the source for a great part of the difficulties encountered in the existing prototype plants.

-These problems are manageable and can be handled by good design including in particular

·fast "first stages" (receiver + energy transport system to storage and e.g., steam generator)
·higher solar multiples and storage
·carefully chosen power conversion systems, matched to the solar specific requirements of the plant as a whole
·larger plants (e.g., $\geqslant 30$ MW(el))
·minimizing plant internal consumption (10% of the annual gross output seems to be a feasible goal for larger plants).

-Site selection is very important. Local meteo conditions must be evaluated carefully. On site measurements of direct normal radiation are necessary before final site selection. Mean values are not sufficient, information concerning the intensity distribution in time is required.

In conclusion, we may say that when we started the design of the present generation of solar thermal plants in the mid-seventies, we thought that we would demonstrate commercial operation on a small scale. We were too optimistic. As a matter of fact, we have been one plant generation further away from commercial operation than we thought at the time. This is the reason why I call existing plants "prototype plants" or "experimental plants" and not "pilot plants" as it is usually done. However, if the lessons learned from the existing experimental plants are incorporated properly into future designs, a satisfactory performance of commercially sized future demonstration plants may be expected now.

1) Statements taken from a lecture given at the 2nd Igls Summer School on Solar Energy 1985, 31.7-9.8.1985 (Papers to be published by ESA)

1.2 Introduction to the SSPS Project

One objective of the International Energy Agency's (IEA's) energy
research, development, and demonstration (R&D) program is to promote
the development and application of new and improved energy technologies
which could potentially make a significant contribution to our energy
needs. Towards this objective, the IEA has established and conducted
energy research, development, and demonstration projects, one of which
is the Small Solar Power Systems (SSPS) project built in the province
of Almeria, Spain. This project, performed under the auspices of the
IEA by nine countries (Austria, Belgium, Switzerland, Germany, Spain,
Greece, Italy, Sweden, and United States of America), consisted of the
design, construction, testing, and operation of two dissimilar types
of solar thermal power plants: a distributed collector system (DCS)
and a central receiver system (CRS). They are constructed adjacent to
each other on the Spanish Plataforma Solar in Almeria, southern Spain.
Both have the same rated electrical output (500 kW$_e$ design at equinox noon) and have delivered electric energy to the Spanish grid during
the three-year period 1981 - 1984.

SSPS PLANT

The SSPS plant operation has produced several unique observations.
* Operational experience has been observed with the functioning of a DCS and CRS power plant.

* Different designs of advanced solar technologies (collectors, heliostats, receivers, storage systems) have been tested comparatively as part of a complete power plant system in different operational modes.

* The grid environment of the Plataforma Solar north of Almeria, with statistically the highest solar irradiation of southern European countries is representative of a wide range of future applications of solar power plants.

* The conventional part of the SSPS power plants, which is the power conversion system, has been tested with respect to its viability for solar applications.

The principal objective of the SSPS project was to examine in detail the feasibility of using solar radiation to generate electrical power. In addition, the project had the following objectives:

* Promote cooperation between IEA members in the field of new technologies.

* Demonstrate the technical feasibility of designing and building solar power plants with available hardware.

* Gather operational performance data on such plants.

* Evaluate the viability of the DCS and CRS concepts.

* Design a plant that was optimized to 500 kW_e, but which had the potential for being scaled up or down.

* Consider different geographical applications and operational modes.

* Minimize the investment costs while achieving reasonable operating expenses, good engineering safety, and a long lifetime.

* Assess the further technical development of solar power plants.

The project consisted of two phases: Phase 1 - the erection of the CRS and DCS system, and Phase 2 - test and operation. The project time schedule shows the main events before and during those two years.

Phase	1977				1978				1979				1980				1981				1982				1983				1984			
	I	II	III	IV	I	II	III	IV	I	II	III	IV	I	II	III	IV	I	II	III	IV	I	II	III	IV	I	II	III	IV	I	II	III	IV
SSPS-Specifications trade-offs, feasibility considerations					Stage 1								Stage 2																			
Plant (DCS + CRS) final design (Stage 1)																																
Stage 2 preparations																																
Procurement and installation																																
Plant testing, operation and evaluation																																
Advanced systems tests									Advanced sodium receiver / Increased collector field																				AST heliostats			

SSPS Site Location

This particular site was chosen for its geographical characteristics and because this region of Spain promised favorable conditions relative to the annual amount and intensity of solar insolation.

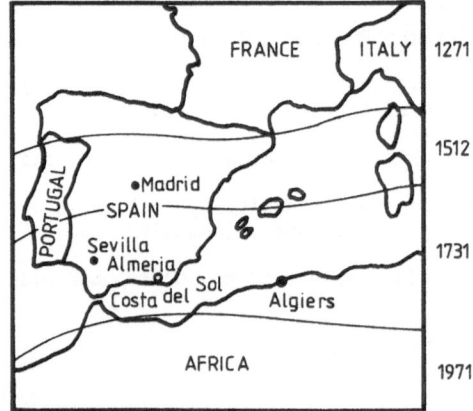

Test and Operation Organization

The testing and operation phase, which was conducted over a period of three years, was organized to collect data on:

- the viability of the selected technical solutions,

- the operational behavior of the systems, and

- the economics of the plants

This phase of the project was administered by the organizational scheme shown below.

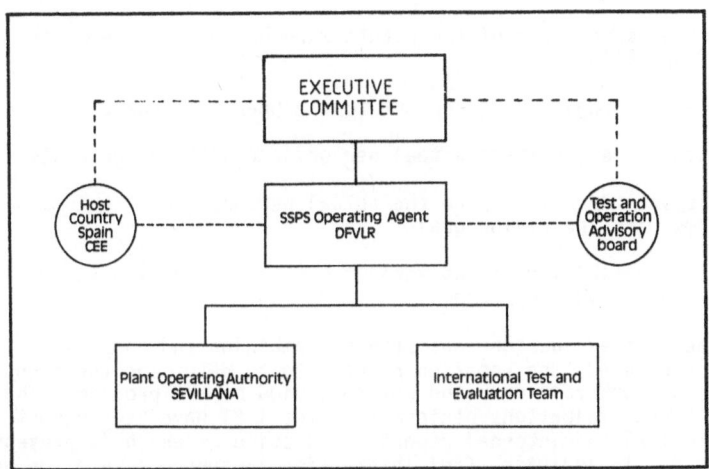

Within this organizational structure, the DFVLR (Deustche Forschungs-und Vasuchsanstalt fur Luft-und Raumfahrt e.V) served as the Operating Agent and was responsible for carrying out the SSPS project on behalf of the SSPS participating countries. The operational and evaluation activities to be performed were specified in a Basic Test and Operation Program document, as well as yearly updates called the Program of Work.
The operation of the SSPS-CRS and -DCS was performed by the regional Spanish utility Compania Sevillana de Electricidad, acting as the Plant Operational Authority.

The scientific testing and evaluation work was entrusted to an international test and evaluation team (ITET) composed of experts from the participating countries that conducted on-site tests and analyses. The ITET was established by the Executive Committee and was headed by Mr. C. S. Selvage. This on-site team has evaluated and reported on test and operation activities and has recommended and advised the plant director on defining, planning, preparing, and conducting tests and operations. The team has performed such functions as:

- recommend tests and modes of operations for the plants

- define criteria to be met for tests and modes of operation and data requirements

- review testing, operations, and maintenance data to assess the validity of the data and potential needs for further data or retesting

- evaluate and report on the results of operations and special tests

- compare the performance of the plants when operating in similar modes of operations

- provide ad-hoc engineering support to the Operating Agent

- at a system level, compare actual performance with design goals

- at a subsystem level, compare the actual performance of the major subsystems with the design goals

- assess the reliability of the various components and subsystems based on an analysis of data.

To summarize, the evaluation consisted of combining and comparing measured, calculated, and reported plant data to determine the plant's performance and behavior over the entire period of the program. The results of these evaluations performed by the ITET have been reported in SSPS technical and internal reports, a listing of which is presented in Appendix A. In addition, four international workshops were conducted on site in order to present the status of the ITET work.

The following are a compilation of new and previously reported studies that represent work done by the ITET performing evaluations of various aspects of both systems that were requested by the TOAB and the Executive Committee. The investigations related to the CRS are described in Volume I of this report; those for the DCS are described in Volume II; and the Site Specific work is in Volume III.

The ITET staff in the years 1981 through 1984 were:

C. Gomes Camacho	Spain	June 1981 - June 1982
A. Baker	USA	summer 1981
R. Stromberg	USA	summer 1981
W. Wilson	USA	summer 1981
M. Loosme	Sweden	September 1981 - June 1983
P. Wattiez	Belgium	September 1981 - December 1984
T. von Steenberghe	Belgium	September 1981 - August 1983
F. Gaus	Germany	December 1981 - December 1983
C. S. Selvage	USA	January 1982 - March 1985
P. Toggweiler	Switzerland	January 1982 - August 1982
H. Jacobs	Germany	February 1982 - March 1985

R. Carmona	Spain	July 1982 - March 1985
M. Pescatore	Switzerland	July 1982 - May 1984
M. Andersson	Sweden	January 1983 - December 1984
J. Martin	USA	May 1983 - December 1984
F. Palumbo	Italy	May 1983 - November 1983
M. Blanco	Spain	January 1984 - March 1985
M. Sanchez	Spain	January 1984 - March 1985
J. Sandgren	Sweden	April 1984 - January 1985
A. De Benedetti	Italy	March 1984 - December 1984
N. Gregory	Switzerland	June 1984 - October 1984
B. W. Swanson	USA	September 1984 - November 1984
W. Schiel	Germany	Part of 1983 and fall 1984
G. Lemperle	Germany	Fall 1984
A. Brinner	Germany	Part of 1983 and 1984

The following is a collection of summaries of each of the final evaluation reports that are published in the three volumes containing the complete reports. The volumes are: Volume I, FINAL EVALUATION REPORT - CENTRAL RECEIVER SYSTEM, Volume II, FINAL EVALUATION REPORT - DISTRIBUTED COLLECTOR SYSTEM and Volume III, FINAL EVALUATION REPORT - SITE SPECIFICS. This collection of summaries is assembled in the same manner as the volumes containing the complete reports with the exception that the Table of Contents represents what is included in the book of summaries, and this book is assembled in series Volume I, then Volume II, and finally Volume III. The introductions to each of the evaluation areas (Sections) are also included in this book of summaries. These introductions contain a summary of each of the specific evaluation reports and the conclusions of that evaluation.

VOLUME I:

CENRAL RECEIVER SYSTEM

CENTRAL RECEIVER SYSTEM

The 500 kW$_e$ CRS plant has a north field of heliostats directing reflected solar energy to a tower mounted receiver. Thermal energy from the receivers is piped to a hot storage tank and then to a steam generator which produces superheated steam. This superheated steam is fed to a steam motor to produce mechanical energy to drive an electric generator. The CRS plant consists of three major systems: a heliostat field, a sodium heat transfer system, and a power conversion system. A simplified process flow diagram of these systems is shown in Figure 1. The CRS main design features are given in Figure 2 and system data in Figure 3.

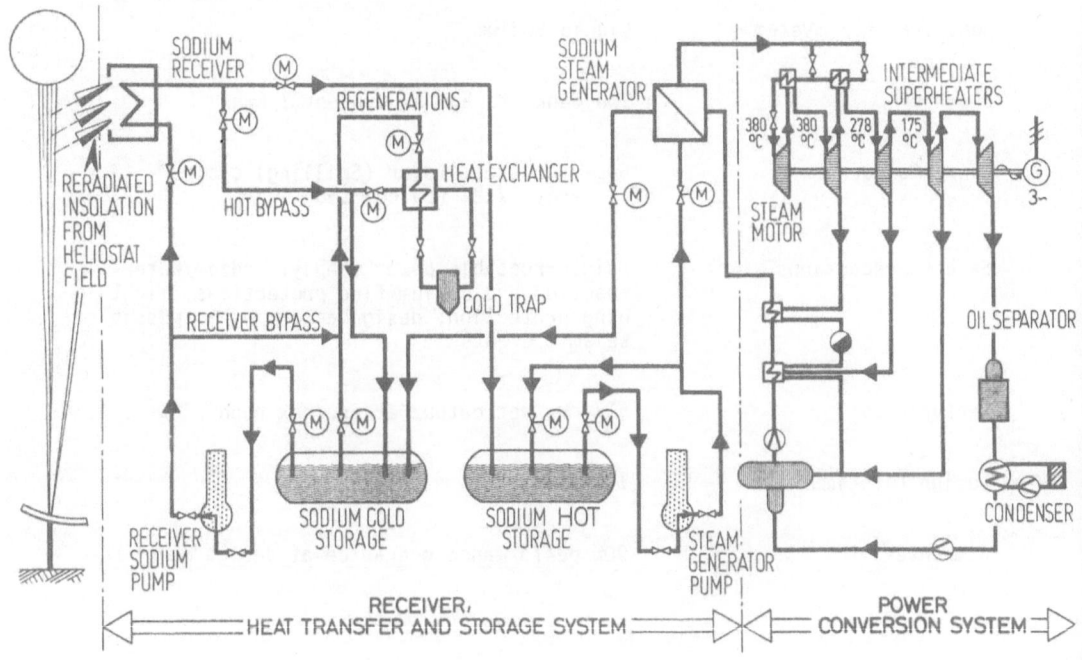

Fig. 1: Simplified CRS Process Flow Diagram

Heliostats	3360 m^2 total reflective area of Martin Marietta Barstow type
Receivers	1) Cavity type with an octagonal shaped aperture of 9.7 m^2; peak heat flux on absorber tubes is 62 W/cm^2; inlet/outlet temperature is 270/530^{0}C
	2) External type with 2.85 x 2.73 m aperture and five panels, each with 39 parallel tubes; peak heat flux of 140 W/cm^2
Heat Transfer System	Liquid Sodium
Storage	Two tank storage equivalent 1 MWh$_e$
Power Conversion	6-piston steam motor (Spilling) cycle efficiency 27.2% (calculated)
Safety Precautions	Uninterruptable power supply, sodium/water-reaction and sodium fire protections, lightning protection, design according to possible seismic events
Performance	517 kW$_e$ net output at equinox noon
Design Lifetime	10 years
Guarantee	90% performance guarantee at design point

Fig. 2: CRS Main System Design Features

Design Point:	Day 80, 1200 (equinox noon)	
	Solar insolation, kW/m^2	0.92

Heliostat Field:	93 heliostats	
	Total reflective surface area, m^2	3660

Receiver:	Aperture size, m^2	9.7
	Active heat transfer surface, m^2	16.9
	Inlet temperature, $^{\circ}$C	270
	Outlet temperature, $^{\circ}$C	530
	Efficiency (calculated), %	85

2nd Receiver:	Aperture size, m^2	7.91
	Active heat transfer surface, m^2	7.91
	Inlet temperature, $^{\circ}$C	270
	Outlet temperature,$^{\circ}$C	530
	Efficiency (calculated), %	94

Thermal Storage:	Storage medium	sodium
	Thermal capacity, MWh	5.5
	Hot storage temperature, $^{\circ}$C	530
	Cold storage temperature, $^{\circ}$C	275

Steam Generator:	Sodium inlet temperature, $^{\circ}$C	525
	Sodium outlet temperature, $^{\circ}$C	275
	Water inlet temperature, $^{\circ}$C	190
	Steam outlet temperature, $^{\circ}$C	510
	Steam pressure, bar	100

Power (at design point):	Solar input to receiver (insolation), kW	2880
	Thermal input to steam motor (thermal), kW	2200
	Gross electric, kW	600
	Net electric, kW	517

Efficiencies (at design point):	Thermal to gross electric, %	27.3
	Thermal to net electric, %	23.5

Fig. 3: Main System Design Data for CRS

The heliostat field subsystem includes the heliostat field and asso-
ciated controls. The Martin Marietta heliostats used are identical to
those at the 10 MW_e Barstow Pilot Plant except that the curvature
of the mirror has been increased to shorten the focal length. Each heliostat has a reflective area of 39.3 m^2. The field consists of
93 heliostats with four different focal zones. Figure 4 gives the heliostat field focal zone definition and shows the concentric-circle
layout of the field north of the receiver tower. The mirror module is
a vented sandwich design of hot-bonded glass mirror, honeycomb core,
and steel pan enclosure. The heliostats are controlled by the heliostat array controller (HAC) located in the main control room. The HAC
transmits commands to the heliostats via four heliostat field control-
lers (HFC). These HFC's act as a heliostat controller (HC) for four
heliostats and also transmits data to other HC's located on each heliostat. All heliostats had the same aim point on the cavity receiver
near the center of the receiver aperture. With the Advanced Sodium
Receiver (ASR), three aiming points were used in order to provide a
balanced thermal input.

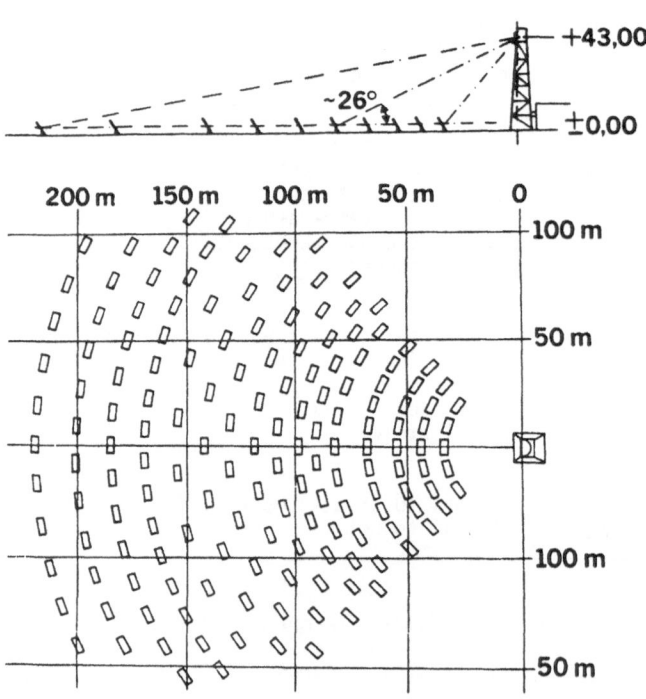

Fig. 4: Heliostaat Field Layout

Two receivers were installed at the CRS. The first receiver was a north-facing cavity type with a vertical octagonal shaped aperture of 9.7 m². The absorber panel is a 120-degree segment of a right circular cylinder. Sodium flows in six horizontal parallel tubes, which are 38 mm in diameter and -1.5 mm wall thickness, which serpentine's from near the bottom of the cavity to the top defining the absorber panel. These tubes are not joined (welded) along their length but are supported by mechanical means. Sodium enters the inlet header at 270°C located at the bottom of the panel and exits the outlet header at 530°C near the panel top. The location of the absorber panel inside the cavity is such that the peak heat flux is about 62 W/cm² at equinox noon when 2880 kW$_{th}$ enters the cavity aperture.

Figure 5: Cavity Receiver (Sulzer)

The second CRS receiver was a 2.7 MW$_{th}$ external type that consists of five panels arranged to form a rectangular absorber 2.85 m high and 2.78 m wide. Each panel consists of a tube bundle with 39 - 14mm diameter vertical tubes, a bottom and top header and a downcomer. The flange of the bottom inlet header and the restraint at the downcomer sodium outlet are attached to the panel. The top header moves vertically to accommodate vertical thermal growth of the panel. The irradiated tubes are assembled together in groups of three and held with four supporting plates to form a 'triplet'. These triplets are connected to the panel framework by means of pins such that the tubes can grow axially with respect to the frame and also rotate, because of the clearance between each pin and its hole in the triplet supporting plate. Gaps are provided between so that each triplet is free to expand independently in the horizontal direction. Liquid sodium is pumped from the cold storage tank at 270°C into the bottom of receiver panel at one edge, through each of the panels in series and out at the top of the central panel at 530°C.

The presence of the gaps between the tubes does allow some concentrated solar flux to pass through the tubes and impinge on the backwall structure. Therefore a double shield of high refractory alumina-based material is located in back of the tube bundle to protect the back structure from the incident radiation. Behind the second shield, a 175-mm layer of ceramic fiber insulates the hot parts of the receiver from the structure.

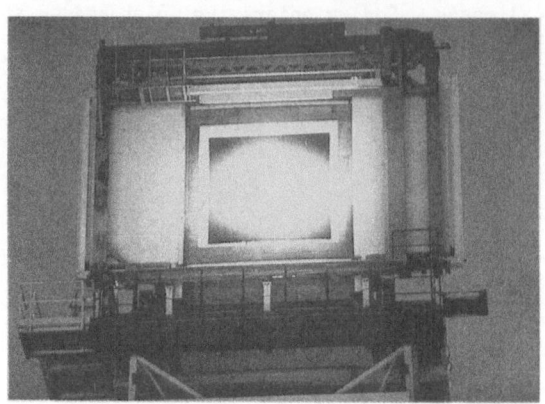

Figure 6: Advanced Sodium Receiver (Franco Tosi)

Storage for the CRS plant is provided by two separate vessels: one cold sodium vessel and one hot sodium vessel. Sodium enters the hot storage vessel at 530°C from the receiver, is drawn from this vessel and pumped to the steam generator, and then is returned to the cold sodium vessel at 275°C. Each of these storage vessels is approximately 3.3 m in diameter, 10 m in length, and has a volume of 70 m^3 and a design pressure of 8.5 bar.

The CRS steam generator is a vertical helical-tube-type with a once-through operation mode. The three heating tubes are coiled around a central displacement chamber filled with nearly stagnant sodium and are housed in a cylindrical shell. Within the tubes, water or steam is flowing from bottom to the top. Hot sodium (525°C) enters the steam generator at the top, flows downward between the outside shell and displacement chamber around the coiled heating tubes where the heat transfer takes place, and leaves through an outlet at the bottom (275°C). Water enters the three helical tubes at the bottom (190°C, 110 bar) and exits as steam at the top (510°C, 100 bar). The nominal thermal capacity of the steam generator is about 2.2 MW.

Figure 7: Steam Generator (Sulzer)

The CRS power conversion unit consists of a steam-driven six-piston reciprocating motor and a three-phase alternating current generator. The steam motor uses intermediate heat exchangers to cool the first stage inlet steam from 510°C to 380°C while heating the second stage inlet steam, the exhaust of the first stage which is at 270°C, up to 380°C. Likewise, the second stage outlet steam, which is at 260°C is heated to 278°C before it enters the third stage. Stages four and five are fed directly from the previous stage without reheat. Extraction connections for steam heated feedwater preheaters are located at the exhaust of stages two and three. Degasification steam is extracted from the fourth stage. The steam motor is elastically connected to a brushless self-controlled threephase current generator equipped with an automatic voltage control regulator, for isolated (non-parallel or stand alone) operation and reactive current control for parallel operation. The operating conditions of the power conversion unit (PCU) are:

Thermal input (steam)	2200kW$_{th}$
Inlet pressure	100-102 bars abs
Inlet temperature	500-520°C
Back pressure	.3 bar abs
Speed	1000 rpm
Motor	845 Hp
Gross electric output	600 kW
Net electricity output	562 kW
Efficiency (gross/thermal)	27.3%
Efficiency (net/thermal)	25.5%

HISTORICAL ASSESSMENT OF SSPS--CRS PLANT PERFORMANCE

INTRODUCTION

The following specific evaluation reports address the actual performance of the CRS through the years 1981 - 1984. The performance is
reviewed, analyzed and then summarized, using the operators daily
and monthly logbooks, the daily/monthly meteo reports, and the data
tapes that were available and usable. The report, PLANT HISTORY, by
N. Gregory, P. Wattiez and M. Blanco, documents this evaluation and
lists outage statistics, major causes of outage, weather and operation data, and energy collected and delivered. It also presents a
discussion of the time taken to heat up the receivers to operating
temperature, based on a statistical analysis of the recorded operational data. The consideration of operational procedures, which have
a dramatic effect on the system heat up time, was considered late in
the evaluation, causing some modification in the conclusions. Consequently, reading this report and comparing this evaluation to the
specific evaluations of receiver performance, which is topic Section
5, RECEVER BEHAVIOR, brings out the differences between what a subsystem can do and what system mismatch and operating procedures allows it
to do.

The conclusions of this report are:

1 - The CRS plant reliability has improved with time. The outage
    percentage for 1984 was considerably less than for 1981 thru
    1983.

2 - The heliostat field system is very susceptible to serious
    damage from lightning strikes with subsequent long outages.

3 - The data acquisition system seems very difficult. Data
    processing from the DAS is unreliable.

4 - Statistically, fewer good days have been lost than bad days.

5 - Operation of the plant on weekends in 1984 could have reduced
    the calculated outage by as much as 20%.

The preceding report stimulated an indepth analysis of performance leading to the next report, DAILY CHARACTERISTICS, wherein M. Andersson and J. Sandgren examined what happens with the CRS on what is defined as a "good day" and how the system handles cloud passage during operation when operating with good day solar conditions. The extensive graphical presentations in this report provide an easy method of learning how the plant operates under these conditions.

The conclusions from this evaluation are:

1 - There is a clear linear relation of calculated energy on the receiver versus energy absorbed by the sodium for the two receivers.

2 - Receiver efficiencies are: 72% for the cavity and 88% for the external,(daily average for a "good"day)

3 - The maximum calculated daily average value of thermal power delivered to the sodium on a per heliostat basis is: 17kw for the cavity and 21kw for the external.

The boundary between sodium and water is always of concern and can be an interesting interface. A detailed description of the sodium/water heat exchanger, the steam generator, is provided by Mr. S. Amacker from Sulzer Brothers, Ltd., the designer and manufacturer of this heat exchanger, with highlights of the experience with this piece of equipment in the report, STEAM GENERATOR EXPERIENCE. Mr. Amacker includes full load and part load performance curves and observes that the steam generator observed performance follows the calculations very closely. This design is "off the shelf" technology and has presented no surprises during these three years of operation.

The conclusions are:

1 - All specified operating conditions are satisfied.

2 - No problems in operation have been observed.

3 - There have been no structural integrity problems.

4 - The design is very flexible regarding system pressure and power needs.

5 - The load change rate could be improved by filling the central cavity with a gas.

6 - Although no problems were encountered, serious attention must be given to the sodium/water interface.

The final evaluation report on the subject of "Historical Assessment",
is the report by the Plant Operating Authority (POA), as written by
J. Ramos, A. Cuadrado and C. Lopez, CRS OPERATIONAL EXPERIENCE. Operational procedures, as developed by the operational team, are described
with some discussion of the rationale for their development. Specific
problems encountered with subsystems and some of the maintenance problems are discussed.

Conclusions are:

    1 - The heliostat field is very vulnerable to lightning.

    2 - Heliostat corrosion is a concern requiring special stow.

    3 - The external receiver requires venting each operational day.

    4 - The Power Conversion Subsystem is very unreliable.

Summarizing all of these specific evaluation reports, it is clear that a
Central Receiver Solar system can be built from existing designs of both
solar hardware and normal power plant hardware. However, much improvement
can result with rather straightforward changes in the solar specific hardware and a considerable change in the power plant hardware. Many of these
needed changes become obvious from evaluation reports that follow.

SSPS - CRS PLANT HISTORY AND OPERATION (1981 - 1984)

Neil Gregory, Pierre Wattiez, and Manuel Blanco, ITET

SUMMARY

The paper presents an overview of the plant history to date (August 31, 1984) and also ascertains some of the principal operating characteristics for the major components/systems, heliostats, receivers, steam generator, and electrical generator. Details are given of the days of operation, operating hours, and major outages that have occurred at the plant since November 1981. In addition, attempts are made at defining characteristic plant operation under normal operating conditions on days which would be classed as good solar days.

In conclusion, despite the high percentage of plant outages that have arisen, it would appear that:

- the outage percentage of 50% for 1984 is a considerable improvement when compared with the outages in 1981-1983. It appears that, with maturity, the CRS plant reliability is improving.
- statistically, fewer good solar days have been lost than bad days, i.e. the plant has had more than its fair share of good solar days.
- the susceptibility of the HFS to serious damage by lightning strikes, and thus long outages, appears to have been reduced with experience.
- a further reduction in outages by approximately 20% should have been possible if the plant were to have been operated at weekends. In addition, the continuity of such operation would definitely prove to be advantageous.

A far as plant operation is concerned, the average number of hours of receiver operation does not differ significantly between the two receivers, (± 6%). In addition, the average generator operation, particularly on good solar days, is approximately the same, (about 3 hours), for both receiver periods. This would tend to suggest that the different performance of the two receivers does not significantly affect the generator operation.

As far as the "start-up" times are concerned, the average values presented demonstrate how the system has been operated and not how the system could be operated. Indeed, both receivers could be started up in a relatively short time; this could be governed by the sodium mass flow and the available power into the receiver. However, the good performance of some components cannot be utilized if the performance of other subsystems does not match.

DAILY CHARACTERISTICS

Mats Andersson and Jonas Sandgren, ITET

SUMMARY

The CRS plant was equipped with a cavity receiver made by Sulzer from November 1981 until April 1983. In December 1983, the external receiver (the Advanced Sodium Receiver (ASR)) was placed in operation. This change did not lead to any major change in system configuration, though it did have some operational impacts.

This paper describes the operation and performance of the CRS-plant on a daily basis and compares the plant operation with the ASR and the Sulzer receivers. The evaluation is based on days where electrical energy was produced for at least 50 minutes, though some other days are used for portions of the study. For the Sulzer system study, 47 days were used during the operational period. For some portions of this study, only 22 of these days could be used in this study because information on the number of heliostats in track was missing. The ASR system study is based upon 32 days with electrical production from December 10, 1983 to August 31, 1984.

There are two parts in this evaluation: a study of various energies and operation times, and a detailed study of some selected days.

DAILY SUMS

Daily sums of various energies are calculated from five-minute averages stored on data tapes by the data acquisition system. The following relationships were found:
- Review of calculated energy into the cavity versus energy absorbed by sodium show a clear linear relation for the two receivers. The receiver efficiencies are calculated as 72% for the cavity receiver and 88% for the external receiver.
- The maximum daily average value of thermal power delivered to the sodium calculated per heliostat was 17 kW for the cavity receiver and 21 kW for the external reciever.

DAILY CHARACTERISTICS

Detailed plots showing the CRS plant operation during three days are presented covering both receivers. Some indication of the controlability is also shown.

STEAM GENERATOR EXPERIENCES

S.W. Amacker, Sulzer Brothers, Switzerland

SUMMARY

This paper describes the thermal design of the CRS steam generator and compares the calculated thermal performance to the measured thermal performance.

This steam generator, using liquid sodium as the heat source, is a once-through to superheat water/steam in three helical tubes in a flowing liquid sodium pool. The specific rating is: thermal power = 2.22 MW, sodium mass flow = 6.76 Kg/s, sodium inlet temperature = 525°C and outlet = 268.9°C at a pressure of 8.5 bar and a sodium pressure drop of 0.5 bar. On the water/steam side: mass flow = 0.87 Kg/s, water inlet temperature = 193°C, outlet = 500°C, with a steam pressure of 100 bar and pressure drop of 10 bar maximum.

In the operating history of the IEA/SSPS plant, there were no problems with the steam generator. There is good agreement between performance measurements and calculations. However, some differences exist within the evaporator part of the steam generator. Heat transfer was underestimated at the film boiling conditions and overestimated at nucleate boiling. However, the inaccuracies have not affected the transferred power. There is good agreement between the calculated and measured position of the critical heat flux.

The operating experiences with the steam generator have led to the following conclusions:

1. The specified operating conditions are met.
2. No problems regarding operation or structural integrity have occurred.
3. The load change rate might be improved by filling the central cavity with a gas.
4. Transient behavior should be investigated.
5. The design is very flexible regarding system pressure and power needs.
6. Although no problems were encountered, the steam generator is an important component of the system. Serious attention must be paid to the boundary between sodium and water.

CRS OPERATIONAL EXPERIENCE

Juan Ramos, Antonio Cuadrado, and Carlos Lopez, Sevillana

SUMMARY

This paper presents the operational experiences collected by the POA[1] in the operation and maintenance of the central receiver system of the SSPS plant. It describes the nonstandard operational procedures that have been developed to allow more efficient operation of the system, the main problems that have appeared, and the maintenance activities it has required.

Special operation procedures are selective defocussing according to the time of day and wind conditions, and modified STOW position to decrease mirror corrosion. In the receiver loop, the ASR requires daily individual venting of the panels, and consequently there has been a need to modify the temperature of the regeneration vessel. A safe procedure to fill the ASR was also developed. The large thermal inertia of the sodium tanks makes necessary the discharge of hot sodium through the steam generator to reach nominal temperatures.

The main problem with the heliostat field has been the damage caused by lightning. The power failure recovery sequence fails occasionally. Installation of the Advanced Sodium Receiver took four months in 1983. The trace-heating system has been the source of many problems due to insufficient power in some sections and lack of redundancy. A manufacturing defect in the cold sodium tank produced several leaks that needed a costly and lengthy repair. In the PCS[2] the steam motor has suffered repeated breakdowns.

1) POA: Plant Operation Authority

2) PCS: Power Conversion System

HELIOSTAT FIELD PERFORMANCE

INTRODUCTION

The heliostat is the first subsystem in a central receiver solar energy
collection/concentration system; consequently, it is the first sub-system
evaluation report contained in this report of the SSPS Central Receiver
System.The heliostat field was designed and manufactured by the Martin
Marietta Corporation.The original field design consisted of 160 heliostats
with 39.3m² mirrow surface each;however, as a result of a shortage of
money during construction, only 93 heliostats were provided and installed.
This caused a mayor change in the definition of the threshold for operation
of the system and a change in the design point insolation level from 700 w/m2
to 920 w/m2.In addition, this changed the ability of the plant to operate
in all of the original modes of operation. This is addressed, in detail,
in the evaluation of system performance, Section7, and Potential for
Improvements, Section 8.

The first specific evaluation report of this evaluation topic was prepared by J.Ramos and P.Wattiez. With a title, HELIOSTAT FIELD HISTORY
AND STATUS, the report describes all the observed events with this field,
analyzes the affect these events had on system performance, and reports
on the field status. The principle result of this evaluation was to
identify areas where significant improvement could be made in design
and construction of an advanced heliostat field system.

The concluding observations or recommendations are:

1- The control system must be automated and more flexible.

2- The automated control system must be influenced by the
   receiver performance such as output temperature.

3- Recovery from power system failure must be rapid.

4- Lightning protection must be provided.

5- Mirrow corrosion is a major problem, at least for
   the particular mirrow module design used here.

6 - A simple, cost effective, washing system is necessary.

7 - Local maintenance is a must in remote areas.

8 - A beam characterization system would be helpful.

Calculations of the heliostat field performance have been performed by
most central receiver system designers using computer codes such as
Helios, Mirval, Delsol, etc. The calculations result in values and
distribution of energy available at the receiver, but the accuracy of
the result is dependent on knowledge of specific heliostats that are
in operation, reflectivity of the mirror field, atmospheric conditions,
etc. Most central receiver installations have made extensive efforts
to measure the energy and its distribution at the receiver interface
in order to validate the heliostat field calculations and to obtain
realistic input for receiver efficiency determinations. In general,
these efforts have experienced extensive equipment difficulties and
indications of large errors in the data collected, with the result that
most efforts have been abandoned. W. Schiel and G. Lemperle developed
a remote flux measuring system that provides frequent calibration in
the range of the measurement, and, therefore, energy flux levels and
distribution that are accurate within a few percent. This effort is
reported in, MEASUREMENTS AND CALCULATIONS ON HELIOSTAT FIELD PROPER-
TIES OF THE SSPS CENTRAL RECEIVER SYSTEM AT ALMERIA, SPAIN.

The conclusions of this evaluation are:

1 - The measured heliostat field efficiency is in agreement with
    the calculated values.

2 - Heliostat beam quality measurements disagree with calculated
    values because of larger than expected sunshapes.

3 - Circumsolar measurements are important in efforts to measure
    heliostat field performance.

The heliostats installed at the SSPS are essentially the same as the
1818 heliostats that are installed at the central receiver facility
SOLAR ONE, in California. Performance of this much larger field is of
value to SSPS evaluation and can provide some interesting guidelines
in the design of new systems. C. L. Mavis and J. J. Bartel prepared a
summation of the Sandia National Laboratories evaluation of the Martin
Marietta heliostats installed at Solar One. The report, 10MWe SOLAR
THERMAL CENTRAL RECEIVER PILOT PLANT - HELIOSTAT EVALUATION, is included here to support the ITET evaluation.

The conclusions are:

1 - In 1983, heliostat availability (1818 heliostats) was 94 to
    99%.

2 - Maintenance hours for 1983 was 160 manhours/month.

3 - Mirror module vents are being installed to control (reduce) corrosion.

4 - The BCS system has been improved and circumsolar (sunshape) measurements are now being made.

HELIOSTAT HISTORY AND STATUS

Pierre Wattiez, ITET and Juan Ramos, Sevillana

This paper describes the evolution of the SSPS heliostat field subsystem during three years of operation. The status of the field is given by a set of performance characteristics which include reflectivity, mirror corrosion, beam alignment, and availability.

Technical problems which have had major influence on the performance of the heliostat field are described, and a survey is given of the experience gained by the Operating Authority.

Finally, conclusions are drawn from these experiences regarding factors that should be taken into acount in the design of future heliostat fields.

MEASUREMENTS AND CALCULATIONS ON HELIOSTAT FIELD

W. Schiel and G. Lemperle, DFVLR

SUMMARY

The HERMES (Heliostat and Receiver Measuring System) has proved to be well suited for measurements of high solar fluxes from complete heliostat fields. The relevant measured data were heliostat field efficiency, beam quality, and tracking accuracy of the field. The results have shown that the solar subsystem meets the design criteria during clear day steady-state operation.

Measured heliostat field efficiency was in good agreement with values compiled by the HELIOS code. From the evaluation, it can be seen that the center of gravity moves 10 cm east and 15 cm down in the morning to 10 cm west and 10 cm up in the evening. This movement depends on both the positions of the heliostats and actual wind speeds.

The beam quality measurements were in disagreement to those predicted due to the enlarged sunshapes, which are typical for the climatological conditions of the plant location. If unfavorable atmospheric conditions exist, the circumsolar radiation can reduce the intercept factor more than 10%. Not knowing the circumsolar radiation means that there is no possibility to establish an exact energy balance. Therefore, efforts must be made to include the effect of circumsolar radiation on the energy balance. Several approaches are feasible. The simplest one is to measure the direct insolation with a pyrheliometer whose view angle corresponds to the effective acceptance angle of the power plant. This approach leads to an acceptance angle of 1.8° for the CRS-ASR.

10 MWe SOLAR THERMAL CENTRAL RECEIVER PILOT PLANT - HELIOSTAT EVALUATION

C.L. Mavis and J.J. Bartel, Sandia National Laboratories Livermore

SUMMARY

Sandia is responsible for evaluating the heliostats at the 10 MWe Solar Thermal Central Receiver Pilot Plant in Barstow, California. The three evalaution objectives are: (1) characterize heliostat performance, (2) identify areas where heliostat research and development may lead to performance improvements, and (3) evaluate the need for a heliostat beam characterization system in future plants.

During the past 12 months, reports have been published detailing Barstow heliostat experiences, mirror corrosion survey results, and 1982 meteorological data. A 1983 Meteorological Report was published in May 1984.

The beam characterization system (BCS) has been upgraded and a sunshape measurement system has been added. Heliostat mirror cleanliness has been measured at two-week intervals, and the effects of rainwashing and spray rinsing of the mirrors have been determined. Mirror module vents are being installed on almost half of the modules to dry out the water that has accumulated inside them to halt mirror corrosion.

During 1983, 94 to 99 percent of the heliostats were in operation at any one time. Maintenance hours are estimated to be 160 hours per month for 1983. The actual hours have not been determined yet. There were 817 maintenance actions during 1983 as compared with 929 in 1982.

During the next six months, evaluation activities will continue, at reduced level of effort after August 1984. An evaluation report for the two year plant evaluation period will be published late in 1984.

RECEIVER BEHAVIOR

INTRODUCTION

Sodium has very attractive heat transfer characteristics, thus it
is an obvious candidate cooling fluid for a high temperature, high
performance, central receiver system. The SSPS project has operated with two sodium receivers: a cavity type receiver designed by
Interatom, constructed and installed by Sulzer Brothers Ltd., and
an external type receiver designed and manufactured by Franco Tosi
Company. The cavity was installed first and operated from mid 1981
until April 1983, when it was replaced by the external receiver, the
ASR. Even though the design requirements for these two receivers
weredifferent, this project provided the first opportunity to make
observations of a cavity and an external type receiver under similar
operating conditions.

This evaluation area (section) contains detailed descriptions of the
two receivers prepared by the designer/manufacturer, several reports
of specific evaluation of performance and losses, some details of the
methods used in conducting these evaluations, and a few reports which
are attempts to make comparisons. The reader is counseled to treat
these comparisons with care, simply because the two receivers were
designed to very different specifications and for different uses. The
first receiver was required to be a conservative design, because this
was the first use of sodium as a heat transfer fluid in a solar central
receiver system and conservatism was very important, wheras, the external receiver was, at the start, an advanced sodium receiver, the ASR.

The first report in this section is a theoretical introduction to receiver design considerations written by A. De Beneditti and J. Martin,
where they relate the theory to the thermal performance as observed
in operation of the SSPS receivers. They expended considerable effort
during this evaluation to assure that the theoretical development realistically explained receiver performance, particularly the ASR.

The conclusions are:

    1 - Performance analysis of the ASR has confirmed design
            objectives.
    2 - Observed thermal losses are lower than expected.
    3 - A factor of 10 scale-up appears reasonable.

The second report is a detailed discussion of the design, analysis and construction of the external receiver as seen by the Franco Tosi/Agip team of P. Cavalleri, V. Bedogni, A. Di Meglio of Franco Tosi, and A. De Benedetti, and C. Sala of Agip. This report provides a detailed description of the receiver, its instrumentation and operational constraints. The modifications of heliostat pointing and the need for this modification is also discussed. There are no conclusions in the report; however, some recommendations for receiver design are provided.

The third report provides a description of the cavity receiver as constructed and installed by Sulzer Brothers Ltd. The report is prepared by H. W. Fricker who led the team from Sulzer Brothers. Although the intent of this report is to provide a reference description of the receiver, Mr. Fricker has also presented some results of his evaluation of the receiver performance.

His conclusions are:

    1 - The SSPS cavity is a conservative design.
    2 - The SSPS cavity has worked well and is easy to use.
    3 - Storage capacity of the ceramic wall was not sufficient.
    4 - Significant improvement in performance is possible with
            simple modifications.

Having performed his evaluation of the cavity receiver and then later serving in a consultant role during design of the external receiver, Mr. Fricker proceeded to prepare an evaluation report comparing the two receivers. The reader is again cautioned, in reviewing this report, that the two receivers were designed to very different requirements, and that Mr. Fricker's position, as leader of the design team of one of the receivers, should be considered. The report is clearly very helpful, in that it provides some comparative charts.

Conclusions are:

    1 - The SSPS external receiver is more efficient than the SSPS
            cavity receiver.
    2 - Both receivers performed their intended functions.
    3 - The cavity is easier to use than the external.
    4 - Modifications to either receiver could resolve any identified
            problem or constraint.

Performance measurements were made on both of the receivers during normal operation and in very specific measurement efforts referred to as measurement campaigns. G. Lemperle and W. Schiel report on their efforts during one of these measurement campaigns where the cavity receiver was undergoing tests and they were developing new interesting techniques for measuring the energy input to the receiver, providing one of the more important inputs to determine receiver efficiency. This report, EFFICIENCY AND TEMPERATURE MEASUREMENTS, effectively describes the techniques used and the result. The conclusion is that: "The measured data compare rather well with calculation of receiver losses for the cavity."

Continuing with the evaluation of measurements and tests, R. Carmona, H. Jacobs and M. Sanchez provide a summation and evaluation of testing with the receivers in the report RECEIVER LOSSES; RESULTS OF TESTS. This evaluation takes into account the problems of inaccurate data as produced by the flow meter during early testing, and then provides a realistic efficiency comparison of the two receivers. Again, there are well understood reasons for some of the efficiency differences that can easily be changed.

The conclusions of this evaluation are:

1 - The losses with the cavity receiver are higher than for the external receiver. At 350°C these losses are 450kw vs 150kw.

2 - The difference appears to be the difference in convective losses.

After developing a theoretical analysis of receiver performance, A. Benedetti and M. Blanco developed a thermodynamic model of the external receiver (Theresa), then applied this model in a comparison with acquired performance data. This report, SIMULATION AND COMPARISON shows, in a graphical form, how well the model represents the receiver. The only conclusion is that the model does produce results that compare well with the actual receiver performance data.

In the last days of operation with the cavity receiver, in late March and April 1983, a rigorous test program to obtain comprehensive performance data was implemented. This program was specifically intended to obtain transient response data with this receiver and consequently the data recording was requested to be in increments of 30 seconds, normally referred to as "point and class summaries". Unfortunately, the ability to process this type of data was developed rather late in the evaluation period and, therefore, the data had not been analyzed prior to the summer of 1984. Mr. N. Gregory attempted to unravel this data in late 1984 and, with great difficulties, found sufficient data to prepare the report TRANSIENT RESPONSE OF THE SULZER (cavity) RECEIVER.

The conclusions of this evaluation are:

1 - The data tapes with the point and class summaries are difficuilt to work with.
2 - The cavity operates without problems in transient conditions.

With improved data handling capabilities so that transient data became a more common possibility, the external receiver, the ASR, was subjected to an indepth transient analysis by R. Carmona and J. Martin with the title THE ADVANCED SODIUM RECEIVER (external) TRANSIENT RESPONSE.

The conclusions are:

    1 - The ASR can be heated to operating temperature from a warm
        condition in 2 minutes during mid day.
    2 - Early morning start-up requires a longer time because of
        reflected sun shape and the small receiver.

A comparison evaluation performed by the operating staff from the
organization Cia Sevillana De Electricidad, addresses operational
constraints resulting from the particular design charateristics of
the two receivers. As stated above, the design requirements of the
two receivers were very different. Consequently, this comparison
is more a critique of the design specifications than of the receivers,
as they were constructed. The report DIFFERENCES BETWEEN FILLING
STRATEGIES FOR THE SULZER AND ASR RECEIVERS was prepared by F. Ruiz
and A. Cuadrado.

The priniciple conclusion is that the external receiver (ASR) is
difficult to fill, so it is not drained each night. This was never
stated as a design requirement, because the cavity receiver was not
operated that way and was only drained when major shut down was
required.

Looking only at the specific operational data, acquired during "normal
operation," H. Jacobs and C. Selvage performed an evaluation of the two
receivers and reported on this effort in RESULTS ON THE PERFORMANCE OF
THE SULZER CAVITY RECEIVER AND THE FRANCO TOSI EXTERNAL RECEIVER. It
was necessary to apply corrections to the sodium flow data for the early
tests because in the calibration efforts of A. Brinner and F. Reich
(DFVLR Stuttgart), the flow meter was shown to be in error. This flow
meter was replaced prior to the external receiver testing.

The conclusions are:

    1 - The SSPS external receiver is more efficient than the SSPS cavity
        receiver - 91.9% compared to 77.4%.

    2 - The reasons are :

        (a) Higher heat flux in the external receiver
        (b) Smaller size with the external receiver
        (c) Conservative design of the cavity receiver
        (d) Poor heat flux distribution with the cavity receiver

RECEIVER THERMAL PERFORMANCE: THEORY

Alessio De Benedetti, AGIP SpA and José G. Martín, ITET

SUMMARY

This paper discusses topics which are revelant to the quantitative evaluation of the thermal performance of a solar receiver. It outlines the
general problem of thermal balance by considering the heat transport
equation as an element of the receiver wall and describing the difficulties which must be solved or circumvented to arrive at numerical solutions for a real receiver.

Absorption, reradiation, and convection are treated in some detail, bor-
rowing from material which represents the 'state of the art'. For simplified geometries, it is possible to derive analytic expressions for the
radiative efficiency of a receiver, which serves to quantify the radiative losses from cavity and billboard receivers.

To calculate the power absorbed and the power emitted by the receiver active surface, basic absorptance and emittance data must be corrected to
take into account the fact that the surface is not flat. Estimates for
this correction are presented.

A summary of the results reported in the literature on measurements and
analysis of forced and free convection in external and cavity receivers
is included.

An overall iteration scheme is suggested which can be applied generally.
Numerical solutions for cavity and billboard receivers based on this
scheme are presented. These provide a link with the work on simulation
and the results of actual experiments.

RECEIVER DESCRIPTION: BILLBOARD RECEIVER (ASR)

P. Cavalleri, V. Bedogni, and A. DiMeglio, Franco-Tosi Industriale,
and
A. De Benedetti and C. Sala, AGIP SpA

SUMMARY

The ASR receiver was designed and constructed by Franco-Tosi Industriale and AGIP SpA with fundamental contributions from ENEL (Italian National Board of Electricity) for the receiver dynamic analysis and control system design and from CNR (National Research Council), the Italian official representative. The receiver intallation at Almería Central Receiver System power plant was completed in August 1983.

The primary objectives of ASR design were:

- incident heat flux peak density in the range of 100 - 150 W/cm^2

- average incident heat flux in the range of 30 - 50 W/cm^2

- upscaling aspects: the ASR should include basic technological aspects of large future receivers.

The final solution for the selected design was: an external type receiver of 2.7 MWt, consisting of 5 panels arranged to form a rectangular shape absorber of 2.85 m height and 2.78 m width. Liquid sodium from the cold storage tank at 270°C is pumped by a feed pump through the receiver, where the sodium is heated to 530°C. Basic design data are presented in this paper.

Analysis of performance data has confirmed the design objectives with lower losses than anticipated. The design appears amenable to scale-up by at least a factor of 10.

RECEIVER DESCRIPTION: THE SULZER CAVITY RECEIVER

H.W. Fricker, Sulzer Brothers Ltd.

SUMMARY

This paper presents a detailed description of the cavity receiver which
was tested at the SSPS-CRS. The basic layout and thermal analysis were
performed by INTERATOM of Germany. Detailed design, stress analysis,
manufacture, and erection on site was performed by Sulzer Brothers Ltd.
of Switzerland.

The receiver is a box, 6 meters high, 5.8 meters wide, and approximately
6 meters deep. The cavity opening is an octogonal opening, 3 meters wide
and three meters high. The absorbing surface consists of a curved wall
with a mean radius of 2.25 meters, a height of 3.607 meters, and an active absorbing angle of 120°. Six parallel tubes of 38 mm (outside diameter) and 1.5 mm wall thickness, each 87 meters long are held at 4 vertical supports and make 14 horizontal passes across the cavity backwall.
The tubes are held on the east side and are free to expand in the vertical and horizontal directions.

The specific rating of the receiver is: thermal power = 2.8 MW, sodium
mass flow = 7.34 Kg/s, sodium inlet temperature = 270°C, outlet = 530°C
with a pressure drop of 0.45 bar. The design maximum heat flux is 63
kW/m^2.

The cavity receiver was the first sodium-cooled solar receiver of this
size. One of the most important design requirements was the need for
safe, reliable operation. Therefore the heat flux was set at a somewhat
conservative 63 kW/m^2. The result is a design using relatively thick
tube walls, which in turn permitted the use of a small number of relatively large - and long - tubes.

The storage capacity of the ceramic wall was not sufficient to maintain
tube wall tremepratures. The ceramic wall simply acted as a heat shield
behind the tubes.

The cavity reciver performed well, was easy to use, and is a conservative
design. There are many modification possibilities that could improve the
efficiency of this type of receiver.

COMPARISON OF THE TWO RECEIVERS

H. W. FRICKER

SULZER BROTHERS Ltd.

This report is a qualitative comparison of the two receivers. The comparison is made on the bases of: design philisophy, operating results, efficiency and complexity.

Design philisophy: The cavity was the first solar sodium receiver of this size in the world. As a result the basic design was a conservative but practical design, amenable to manufacture and ease of operation. The external receiver, as an advanced sodium receiver (ASR), was designed to press the state of the art, raising the heat flux toward the practical limit of the heliostat field and the sodium pumping system. Both receivers achieved their design goals with a minimum of problems.

Operating results: The cavity receiver has given excellent performance during the initial operation from the spring 1981 kto the end of testing in April 1983. It has operated for over 1300 hours, on approx. 250 days without causing any operational constraints.

The ASR has also given excellent performance once the requirements of filling were understood. It has operated for 171 days, up to August 1984, and has met all of its design requirements. Some operational restrictions are present as a result of the high heat flux design which created a small target and a sensitive side shields. In the early morning and late afternoon, noncircular sun images cause heating of the shields.

Efficiency: Although it is very difficuilt to obtain an exact measurement of receiver efficiency, it seems clear that the ASR has a considerably better efficiency that the cavity. Carefull analysis of the detailed performance of the two receivers helps in understanding how this could be the case.

EFFICIENCY AND TEMPERATURE MEASUREMENTS

G. Lemperle and W. Schiel, DFVLR

SUMMARY

Data collected during the measurement campaign in Fall 1982 were used to determine the thermal efficiency of the Sulzer receiver as a function of power into the cavity, time of day, and beam irradiance. At an insolation of 900 W/m, receiver efficiency of about 87% and thermal efficiency of about 60% were measured. The thermal efficiency over time of operation (8'50") on a clear day (October 7, 1982) comes to 57%. Converting this value to the total day from sunrise to sunset (11'25") by 8'50" of operation, one gets an average efficiency of 50%.

The measured data were compared with a simple relation describing the receiver losses. Although the reliability of this simple equation is not yet proven, good agreement was obtained.

Knowledge of the unsteady temperature distribution on the receiver tubes is a prerequisite for the calculation of thermal stresses and efficiency. Solar central receiver tubes are characterized by a nonuniform heat flux load, especially along the circumference. Using a finite difference scheme, the temperature distribution of the Sulzer receiver has been calculated. The HELIOS computer code was used to calculate the spatial angle dependent concentrated radiation, whih served as an input parameter. Temperature distribution on the tubes, ceramic back wall, in the sodium, and the unsteady response of the sodium temperature to a step function show the thermodynamic behavior of the Sulzer receiver.

To verify numerical simulations, it is necessary to compare them with temperature measurements. Infrared imaging systems combine the advantage of not disturbing the temperature distribution and getting a complete array of the front surface temperature. In the last week of September 1984 an IR-camera was installed, and now some preliminary results of IR-measurements of the ASR receiver are available. To avoid errors by reflected solar radiation, the detector operates in a atmospheric window at a wavelength of 8 to 12 mm. The temperature distribution is presented by isolines and profiles. A maximum temperature of 630°C was reached on the central panel. The profile is compared with the incident heat flux. For this special case, the temperature distribution results in receiver loss due to thermal radiation of about 78 kW. The estimated error of this measurement is ± 10°C, mainly due to a nonuniformity of the emissivity.

RECEIVER LOSSES: RESULTS OF TESTS

Heinz Jacobs, Manuel Sánchez, and Ricardo Carmona, ITET

SUMMARY

The objective is to evaluate the thermal losses from the receiver over the temperature range at which the receiver may operate. To achieve this, tests have been carried out with no concentrated power reaching the receiver and with the receiver doors open. The tests consist of circulating sodium through the receiver at different inlet temperatures and estimating the losses by multiplying the mass flow rate times the enthalpy loss.

Sodium has been circulated through the receiver in the normal direction (i.e., flowing from the cold storage tank to the hot tank) for inlet temperatures between 200 and 300°C. To estimate the losses at higher temperatures, sodium must flow from the hot tank and therefore tests have also been conducted with sodium flowing in the reverse direction.

This paper is divided into two parts. The first refers to tests performed by the ITET on the Advanced Sodium Receiver since April 1984. The second part refers to tests performed on the Sulzer receiver during 1982, by J. Kraabel from Sandia, and by one of the authors of this paper (H. Jacobs) during March and April 1983. These tests were performed using the sodium flowmeter which had been installed originally on the CRS.

From the test results, the total thermal losses of the ASR receiver can be expressed as a function of the sodium mean temperature minus the ambient temperature as: $L(kW) = -60.4 + 0.623\ T(°C)$, in the interval $170 < T < 400°C$.

The convective losses have been calculated by subtracting from the total losses the estimated losses due to radiation and the losses due to conduction. From work reported by Sandia and by the ASR manufacturers, the conduction losses amount to about 5% of the total loss. The correlation found for the convective losses is: $L_c\ (kW) = -21.6 + 0.291\ T\ (°C)$, in the interval $170 < T < 400$.

From the tests carried out in the Sulzer receiver, the total losses have been estimated as: $L_t(kW) = -253 + 1.93\ T(°C)$, in the interval $170 < T < 400°C$.

ASR PERFORMANCES: COMPARISON WITH SIMULATION

Alessio De Benedetti, AGIP SpA, and Manuel Blanco, ITET

SUMMARY

In this paper the interface between optics (incident flux distributions) and thermodynamics (temperature distributions) is analyzed and the ASR thermal performance is evaluated. The thermal analysis of the receiver has been performed with the thermodynamic model THERESA.

Temperatures in the receiver tubes, headers, and ceramic backwall are computed with the ASR actual geometrical and physical characteristics; the incident flux distributions, the inlet temperature, and the flowrate used as input are equal to measured values.

THERESA can compute temperatures in fluid, tubes, and backwall in both steady-state and transient conditions.

First THERESA computes the power incident on the billboard receiver, establishing how this power is divided between the five panels. Then the incident power distributions on fifteen receiver tubes is computed (the selected tubes correspond to the instrumented tubes) and the thermal balance equations in the fluid, tube wall, and backwall are solved.

Simulation and experimental results are compared under steady-state conditions (full or partial load operation, loss tests, transient conditions, start-up, input power variations, pump blockage) and special situations (higher flux and temperature, etc).

TRANSIENT RESPONSE OF THE SULZER RECEIVER

Neil Gregory, EIR

SUMMARY

During the period March - April 1983, many tests relevant to the Sulzer (cavity) receiver were performed, including some cloud tests which should have provided a source of suitable transient data.

In addition to the daily 'five-minute average' data, additional data for these test days were recorded on three 'point and class summary tapes' by the data acquisition system (DAS) for later evaluation.

Many difficulties were encountered in reading these tapes and unfortunately success was limited to a single day, April 6, 1983. Further analysis of this day has revealed that although no cloud test had in fact been performed, the transients which occurred were the result of a grid failure and a subsequent receiver trip (emergency shutdown operation).

This paper contains a discussion of the event which lead to these transients and also a description of their major features.

THE SSPS ADVANCED SODIUM RECEIVER: TRANSIENT RESPONSE

Ricardo Carmona and José G. Martín, ITET

SUMMARY

High-flux sodium-cooled receivers have low thermal inertia and short response times. Controls must keep the outlet temperature constant at
changing power levels, without compromising structural integrity. The
Advanced Sodium Receiver (ASR) and its controls are designed to accom-
plish this; it is of interest to determine how close performance approaches expectations.

The ASR transient response has been tested by making changes in the sodium flow rate and the heat flux reaching the receiver at different
loads. From the tests, one estimates gains and time constants. These
vary inversely with the Na-mass flow W (in kg/sec) and are close to computer predictions. The receiver time constant in seconds is about 250 divided by W(in kg/sec) ; the gain (Na-output temperature increase in $^\circ$C
per kW of solar radiator flux increase) is about 0.73 divided by W.

The receiver configuration and actual response suggest a form for its
transfer function. The coefficients have been identified: the temporal
response evaluated from this function fits the data well. It is now possible to predict the response to arbitrary changes in the operating conditions, and to conduct system and stability studies.

This paper presents some consideratiuns on the time required for warm
start-up. The receiver can be heated from 270 to 530°C in about two
minutes at noon. For early morning warm start-up, the time required is
several times longer.

Finally, the paper presents observations on the performance of the receiver loop (i.e., the system composed of receiver, pump, and controls) when
clouds pass over the heliostat field. The controls do protect the receiver from overheating when the insolation rises sharply, but it io desirable to improve the accuracy of the irradiation signal.

DIFFERENCES BETWEEN FILING STRATEGIES FOR THE SULZER AND ASR RECEIVERS

Francisco Ruíz and Antonio Cuadrado, Sevillana

SUMMARY

In this paper a comparison is made of the procedures used to fill with sodium (cold start) of tne two receivers tested at the SSPS-CRS.

The filling of the cavity receiver (Sulzer/INTERATOM) is a straightforward procedure due to the simple path that the sodium must follow. It only requires preheating of the tube bundle and the combined action of the sodium pump and pressure difference between the cold tank and the regeneration vessel.

On the contrary, the more complicated structure of the external receiver (Franco-Tosi/AGIP) makes necessary a carefully controlled procedure based on pressure differences alone, with parallel filling of all the receiver panels through the inlet and draining pipes. The uniform flux distribution required for preheating is obtained by changing the aiming point of some selected heliostats to cover all the receiver area. The backwall temperature has to be raised to help equalize the temperature distribution in the receiver panels.

The difficulty of the ASR filling procedure penalizes operational cost, since the need to keep the sodium flow through the receiver 24 hours a day requires the continuous presence of one trained operator, even at night and during weekends.

RESULTS ON THE PERFORMANCE OF THE SULZER CAVITY RECEIVER
AND THE FRANCO-TOSI EXTERNAL RECEIVER

Heinz Jacobs and Clifford S. Selvage, ITET

SUMMARY

This paper presents some of the results of evaluation performed on the
cavity receiver (Sulzer) and the external receiver (ASR) installed at the
IEA/SSPS project. These two different receiver concepts were tested in
the 500 kW central receiver system experiment at the Small Solar Power
Systems Project in Almería, Spain. Both receivers use liquid sodium as
the receiver coolant or heat transfer fluid. The system is storage
coupled with the sodium heated by the receiver, flowing to the storage
vessel, then to the power conversion system.

Data from operational days with both receivers is used in this evaluation
as an aid in determining receiver performance, losses, and efficiency.
Tests performed specifically to measure losses were conducted on both
receivers, and the data acquired supported the conclusions reached by
operational data analysis.

The conclusion is that the Advanced Sodium Receiver (ASR), the external
receiver is more efficient than the earlier designed and tested cavity
receiver. Some reasons for the differnces as discussed in the paper are:

        1. Higher heat flux on the ASR
        2. Smaller size of the ASR
        3. Conservative design of the cavity
        4. Poor heat flud distribution in the cavity

THERMAL LOSSES/THERMAL INERTIA

INTRODUCTION

Quantifying the loss stair-step has been one of the important goals
of the evaluation team at the SSPS project. This evaluation area
addresses each of the loss elements as they were evaluated, and also
discusses one of the interesting realizations that resulted from this
evaluation - thermal inertia, a major hinderance to successful solar
thermal systems application.

H. Jacobs and M. Andersson analyzed the different operating modes of
the over-all system and evaluated the piping and storage tank losses
over the various operating modes. The report on this evaluation is
in the report LOSSES OF THE PIPING AND TANKS. The normal cool down
charateristics of the hot and the cold sodium storage tanks is presented. The energy loss, when sodium is kept flowing through the
receiver during non solar hours, is also discussed. We are introduced
to the subject of thermal inertia.

The conclusions are:

    1 - Losses and thermal inertia in the pipes and tanks limit plant
            operation.
    2 - The difference in normal receiver outlet temperature and the
            PCS operating temperature is small, therefore losses in the
            storage tanks are very important.

Looking next at the Power Conversion System (PCS), the losses due to
low sodium temperature, start-up losses, and losses during normal
operation, are evaluated by H. Jacobs and R. Carmona and reported on
in the document PCS LOSSES.

The conclusions of this report are:

    1 - The Power Conversion System requires more energy than the
            rest of the system can collect.
    2 - The amount of thermal losses observed is a function of the
            time since the PCS was last operated.

Although it would seem appropriate for a report on the receiver losses
to be included in the preceding evaluation area, Receiver Performance,
the report REMARKS ON RECEIVER BEHAVIOR is included in the loss evaluation area in order to include all loss areas in one section. C. Selvage
used this report to formalize several internal reports addressing the
area of receiver losses. Work by several previous ITET members where
receiver absorptivity, reflectivity and conductivity were measured and
reported on, are referenced in this report with each referenced report
duplicated for future reference.

The conclusions are:

    1 - Each of the loss elements in the system are important and
          most have been measured.
    2 - The accuracy of these measurements could be improved.

A very serious loss area, that remains a serious problem, is the requirement to trace heat to all of the containment metal for the sodium.
This is not a problem unique to a sodium cooled system but any system
whose coolant solidifies at temperatures above ambient, such as molten
salt. P. Wattiez and A. Cuadrado surveyed the electrical power consumed
by the SSPS/CRS trace-heating system and provided the report CRS PARA-
SITIC CONSUMPTION: THE TRACE HEATING.

The conclusions are:

    1 - Trace heating uses much of the energy generated.
    2 - Trace heating of the SSPS plant can be reduced by careful
          plant operational management.

The intermittency of the solar energy source makes imperative the
consideration of time dependent effects in the evaluation of solar
thermal plants. At the SSPS project, the "loss stair-step" concept,
which guided most of the early evaluation work, has gradually been
complemented by an "inertial ladder" concept in which different subcomponents have characteristic time responses and delays which have
an important effect on the actual plant performance. This "paradigm
shift" is discussed in some detail by C. Selvage in the report
IMPLICATIONS FOR DESIGN AND OPERATION.

The conclusions are:

    1 - Thermal inertia is a major problem in the SSPS/CRS.
    2 - Thermal inertia can be designed around.

PIPING AND TANK LOSSES

Heinz Jacobs and Mats Andersson, ITET

SUMMARY

In this paper, the thermal losses for the main piping and tanks in the CRS-system are presented. The main objective is to provide the necessary data for calculating the overall losses in the system at normal operation conditions. The calculations are based mainly on data from the CRS-DAS (Data Acquisition System) collected during the period May to September 1984, after the calibration phase which was finished in the end of April.

The losses in the main piping of the CRS-system have been calculated under different operational conditions. These losses vary from 0.7 to 1.2 W/m * K between different pipes. The large number of supports, hangers, and shock absorbers has a large influence on the loss figures.

The losses of the two storage tanks have been investigated by observing the cooling-down behavior. The evaluated heat transfer value is 65 W/K for the cold storage and 45 W/K for the hot storage.

CONCLUSIONS

Losses and inertia in pipes and tanks limit the possibilities of the plant operation. Since the plant must be put into operation every morning, the thermal energy used to heat-up the piping has a large effect on the start-up time.

Due to the fact that the difference between the nominal receiver outlet temperature and the temperature required for the PCS is small, the loss figure for the storage is very important. A small drop in temperature of the hot sodium is enough to make it useless for steam production.

CRS POWER CONVERSION SYSTEM LOSSES

Ricardo Carmona and Heinz Jacobs, ITET

SUMMARY

This paper presents a short discussion of the performance and results of
evaluation performed on the power conversion system (PCS) of the IEA/SSPS
central receiver system (CRS).

The PCS consists of a sodium steam generator, heat exchangers, condensor,
cooling system, and a five-stage steam engine connected to a 600 kW alternator. How the plant works and the regulating system is built are described in this paper.

Evaluation based on data received during operational days produces the
following results:

- The start-up of the PCS needs thermal energy -- approximately 20% of
  the daily energy offered by the sodium is used for this. It is well
  known that this figure can be reduced by having longer daily operational times.

-During alternator operation, the efficiency is lower than 30%. The actual figure depends on the load. The efficiency under nominal conditions
  (599 kW(el),gross) is 21%.

The conclusion is that the PCS for the SSPS plant is too large or in
other words that the solar multiple is too small.

A higher solar multiple and a PCS with lower power needs would result in
higher total efficiency. The thermal losses and the start-up energy
needs could be drastically reduced.

REMARKS ON RECEIVER LOSSES

Clifford S. Selvage, ITET

SUMMARY

This is a survey paper with the intent of addressing the loss elements of
a receiver and identification of the specific efforts taken by ITET members and associates to measure these losses. The internal reports that
discuss these efforts, the technique, the data, and the analysis results
are included, and therefore provide documentation of these efforts.

The loss elements are seperately discussed and the methods used to measure them are described. These elements are:

     1. Reradiation
       absorptivity - emissivity
     2. Convection
     3. Conduction

Not all of the efforts attempted were successful. Not all of the desired
data was acquired. However, the most significant losses were measured in
several ways and the results support other analysis as reported in this
book.

CRS PARASITIC CONSUMPTION: THE TRACE-HEATING

Antonio Cuadrado, Sevillana and Pierre Wattiez, ITET

SUMMARY

One of the first priorities when selecting the operation strategy for a solar power plant built to produce electrical energy is to reduce the plant internal electrical consumption to a minimum. This paper presents the results of a specific survey of the trace-heating electrical consumption during the first seven months of 1984.

The trace-heating consumption represented 60% of the total consumption for the CRS during an operation day and 72% during a nonoperation day. After the modifications and changes in the operational strategy the internal consumption of the trace-heating was 41% on the fourth day of operation and 77% on the fourth day of nonoperation.

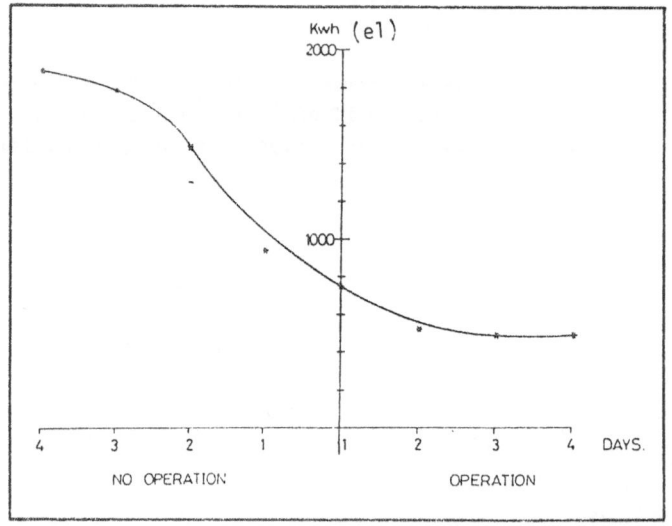

CRS Trace-heating Consumption versus SHTS Operation History

One can state that trace-heating consumption is heavily dependent on the preceeding days' operational pattern (thermal memory). The curve obtained can be helpful in the estimation of long-term trace-heating consumption if the stastistical distribution of operational days is known. Based upon these data, careful management of operation can produce substantial savings in traceheating consumption. Using thermal energy instead of electricity for traceheating should be considered.

IMPLICATION FOR DESIGN AND OPERATION

Clifford S. Selvage, ITET

SUMMARY

Thermal inertia is present in all thermal systems and in large power conversion systems, such as fossil-fired steam/electric generating plants it is
beneficial. This is simply because the plants are started very slowly,
then kept running over long periods of time and the thermal inertia reduces thermal shock.

Solar thermal systems cannot be continuous in operation because the energy source, the sun, provides us with intermittent energy.

This paper addresses the problem of thermal inertia, briefly discusses
the addition of thermal mass to one of the receivers tested at SSPS and
the effect of that mass, then introduces a mathematical consideration of
the thermal inertia of SSPS-CRS. All of this leads to the realization
that this thermal effect is real and cannot be eliminated, but can be
accomodated by proper consideration of the energy losses and procurement
of thermal to electric conversion equipment that can function within this
larger than normal temperature variation.

SYSTEMS ASPECTS/CONTROL

INTRODUCTION

The favorable heat transfer characteristics of sodium makes it possible
to collect concentrated solar radiation at medium high temperatures in
small active areas, and therefore, with relatively low losses. However,
because of the intermittent nature of the solar source, the receiver and
the sodium pump controls must be designed and operated so as to obtain
adequate responses to insure receiver structural integrity. The two
receivers installed and tested at the site represent two different coolant flow approaches, each of which pose different demands on the control
system. This evaluation area reports on studies of this problem situation
and in the final analysis, attempts to make clear the necessitity of system integration; that is, the need to design a coolant flow control system
specifically for a receiver. This was not the completly true case at SSPS
for either receiver. Even with the cavity installation the control was
designed for an input power level much higher than was finally realized.
The ASR, of course, was required to accommodate the coolant flow/pumping
situation that was installed with minimum modification.

In a comprehensive review of the general problem of temperature control
of central receivers, C. Maffezzoni, in the report TEMPERATURE REGULATION;
CRS, discusses the process dynamics for different types of receivers, possible control approaches and practical control schemes for receivers cooled
by sodium. This is a rather theoretical analysis which leads the reader
through the theory and to the conclusions.

The conclusions are:

    1 - Traditional temperature control concepts are unacceptable
        for fast responsive solar systems.
    2 - Newer concepts are workable.

The cavity (Sulzer) receiver was designed with a heliostat field
of 160 heliostats of 39.4 m2 each. As a result of insufficient
funds, only 93 heliostats of 39.4 m2 each were installed, thus dra-
matically reducing the maximum possible energy input to the receiver.
D.Weyers does not dwell on this fact in the report TEMPERATURE
REGULATION, but a reasonable description of the control system
is provided along with some observations of the control system's
performance.

The conclusions are:

    1 - The traditional feedback control system, used with the cavity
        receiver worked rather well.

    2 - The prime control variable was outlet temperature.

The conclusions are:

    1 - The traditional feedback control system, used with the cavity
        receiver worked rather well.
    2 - The prime control variable was outlet temperature.

The external receiver designers were required to interface with the
coolant pumping system and pump control system that was designed for
the original receiver/heliostat field system. This receiver was also
designed to be very responsive to solar energy input variations, which,
with the constraint of given pumping capabilities, required the design
and implementation of a very sophisticated control system. G. Magnani
describes this control system in the report TEMPERATURE REGULATION; ASR.

The conclusions are:

    1 - Feedforward control was essential for the fast response
        receiver, the ASR.
    2 - The feedforward control system eliminated the concern for
        rapid increases in absorbed power.

Where previous control studies and designs approached the control
problem in a responsive mode, M. Blanco and M. Sanchez address the
control problem by controlling the energy input. Their efforts are
described in the report TRACKING; CONTROL OF INCIDENT POWER AT THE
RECEIVER. Following the development of a theoretical method, a test
of this approach was made at SSPS and that effort is included in the
report.

The conclusions are:

    1 - A computer program is developed to control the heliostat
        field in response to receiver energy needs.
    2 - Experiments demonstrated that the system can be made to
        work.

In an attempt to bring "lessons learned" into this evaluation effort,
C. Selvage comments on the positive aspects of various solar thermal
electric approaches. Considering the experiences with sodium as a
heat transfer fluid and its limitations as an energy storage medium,
Selvage suggests the value of a sodium/salt system.

The conclusions are:

    1 - Sodium is an excellent heat transfer medium and has
        demonstrated its usefulness as a receiver coolant.
    2 - Molten salt has a high thermal capacity which is necessary
        for thermal storage.

TEMPERATURE CONTROL OF SOLAR RECEIVERS

Claudio Maffezzoni, ENEL

SUMMARY

The effective use of new energy processes essentially relies on the possibility of ensuring operation continuity to maximize the integral energy
output. This is particularly true in the case of central receiver solar
plants where a reasonable return of the capital investment requires daily
production periods to be maximized, in spite of severe environmental disturbances like those due to clouds passing over the mirror field. Morev-
er, optimization of receiver efficiency may lead to the adoption of very
high thermal fluxes on the receiver surface, which implies very high temperature deriviatives and thermal stresses in connection with solar disturbances. These loads call for receiver temperature control systems
with very good response times to keep the temperature variations within
acceptable limits during fast cloud passages.

This paper demonstrates that traditional temperature control concepts are
unacceptable in that they are generally not designed for very fast feedback control. The reason for this deficiency is essentially because receiver dynamics is affected by relevant transport delays. These delays
prevent single temperature control loops from the receiver outlet temperature to the coolant flow demand from being effectively closed.

A critical analysis of traditional control concepts is performed to point
out that a new control concept can effectively be applied to high temperature receivers. The new concept proposed is based on the feedback of
temperature measurements taken at intermediate receiver points.

The paper addresses: the dynamics of sodium-cooled receivers and its re-
lations to the achievable control performance, the extension of dynamic
analysis results to different kinds of receivers (e.g., gas- and water-
cooled receivers), the basic principles for the control of sodium receivers, and describes the practical realization of a new control scheme
(PFTC)[)] for sodium receivers. Finally, the extension of the control design principles to different types of receivers is discussed.

1) PFTC: Predictive Feedback Temperature Control

SULZER FEEDBACK CONTROL CONCEPT

Dieter Weyers, INTERATOM

SUMMARY

This paper describes the temperature control system for the Sulzer cavity receiver as operated at the IEA/SSPS-CRSA from September 1981 to April 1984.

This system is a feedback control system to adjust the coolant flow in order to maintain a constant 530°C outlet temperature consistent with a lower flow limit of 10% of maximum flow, and protection of the receiver tube bundle in case of tube overheating.

The prime control variable is outlet temperature. This control variable is overridden by tube bundle temperature if that temperature gets too high and by minimum flow to prevent too low a flow. A sun presence sensor serves as a disturbance variable to detect rapid changes of sun energy.

THE ALMERIA ADVANCED SODIUM RECEIVER: DYNAMIC ANALYSIS AND CONTROL

G.A. Magnani, SdI

SUMMARY

The Advanced Sodium Receiver (ASR) of the IEA/SSPS Almería plant is an uncommon heat exchanger because of its high thermal fluxes (138 W/cm^2 peak at the design point) and because the absorbed power is not controllable and is variable from its maximum value to zero and viceversa in a few seconds.

It is clear that the dynamic behavior of receiver temperatures plays a crucial role in the useful life of the receiver; therefore transients are an important consideration in the structural design of the receiver.

This paper deals first with ASR distributed parameter dynamic analysis and simulation, and then it describes the ASR control system based on the new concept of Predictive Feedback Temperature Control (PFTC), together with the additional practical expedients necessary to cope with full load variations due to cloud passages. Finally the direct digital controller is briefly described, and some experimental results are shown to validate modelling and to evaluate control system performances.

The accurate distributed parameter mathematical modelling of the ASR has led to a good understanding of the temperature behavior related to both the sodium flowrate and absorbed power variations.

The simulation of cloud passage transients has shown a relative weakness in the feedback control action where the absorbed power increases rapidly from the minimum load (10%). However a very rough (50% error) feedforward action allows the controlled system to function well even in these worst cases.

Simulation and stress analysis have shown that the most damaging transients to the ASR are due to rapid increases in the absorbed power following a time of heliostat field shadowing. In this case the receiver temperatures are lower than the nominal profile and high temperature gradients arise even with "perfect" control. The control algorithm has been improved to minimize these gradients, and structural design has been modified so that the ASR can withstand them.

CONTROL OF INCIDENT POWER AT THE RECEIVER

Manuel Sánchez and Manuel Blanco, ITET

SUMMARY

This paper describes a computer program which has been developed to control the power sent to a receiver. The program, 'CONTROL', takes heliostats in and out of track in 'real time' to achieve this goal.

To control the power, one must know the useful power which each heliostat
can send to the receiver at any given instant. Based on this characterization of the heliostats, a program has been prepared which identifies
the heliostats which must be put in track to achieve a desired power
level ('set-point') at any solar time, given the solar irradiance.

The preparation of this program was based on the following concepts:
(1) Heliostat Status
    Each heliostat is classified according to the power sent to the re-
    ceiver.
(2) Heliostat Motion
    This refers to the program output option which identifies the change,
    if any, of a heliostat from 'tracking' to 'standby' or viceversa.

The decision to move the heliostats is based on the criteria that the
number of heliostats whose status is going to be changed be as small as
possible. To ensure this, the heliostats in the 'tracking' and 'standby'
sets are arranged in the order of decreasing useful power. The program
will choose alternately, heliostats with high or low useful energy, as a
funcion of the difference between the set-point and the power sent to the
receiver at the earlier 'instant', until the error which is made when
'moving a heliostat' is larger than the error made if that heliostat is
not 'moved'.

'CONTROL' has been implemented on the VAX 11/730. The CPU time for the
main loop is about two minutes in this minicomputer.

The power sent to the receiver was calculated for October 10, 1984 from
the five-minute iradiance DAS readings from 11:00:00 until 14:30:00. The
relative errors between the set point and the estimated power sent to the
receiver are less than 1% except in two instances when the set point was
less than the power contributed by the set of heliostats, which must
necessarily be in track.

POTENTIAL FOR IMPROVEMENTS

INTRODUCTION

This evaluation area (section) addresses one of the original objectives
of the SSPS project -- the possibilities of this technology for the
future. Certainly all of the evaluation reported on in this volume
contributes to developing possibilities for improvement of the Central
Receiver System. Each of the shortcomings of the SSPS system led to
improvement suggestions so, as these were identified, modifications were
indicated, and even small disappointments have led to suggestions for
improvements.

An example of an effort in one area which led to improvement ideas
resulted from a major calibration effort performed by A. Brinner from
the DFVLR, Stuttgart, and led Brinner and W. Schiel to prepare the
report, IMPROVEMENTS IN MEASUREMENT EQUIPMENT. In preparing this report
they drew on experience from the SSPS - CRS to make concrete suggestions
regarding measurements in future experimental power plants.

Their conclusions are:

    1 - Measurement sensors should be calibrated frequently.
    2 - More measurement points are needed.
    3 - Control and measurement systems should be separated.

Reviewing the performance data from all of the evaluations reported in
this volume, as well as some SSPS/ITET internal reports, J. Martin and
R. Carmona suggest improvements which are practical with the SSPS - CRS,
in POTENTIAL FOR IMPROVEMENTS. In this report they make estimates of the
effect these suggested improvements could have on the overall efficiency
and the reliability of the plant, presenting these effects in a graphical
form. Further, they make a distinction between changes on the present
plant at a reasonable cost and modifications which might affect the design
of future and larger plants.

Their conclusions are:

1 - Small efficiency improvements are possible by frequent
      heliostat washing, heliostat tracking and checks of beam
      quality.
2 - Major improvements are possible by changing the PCS and
      redesigning the heliostat field layout - perhaps to 20%
3 - Plant loads (parasitics) must be reduced.
4 - Sodium is not a good storage medium.

In an attempt to bring "lessons learned" into this evaluation effort, and
to make use of these experiences, C. Selvage comments on the positive aspects
of the various solar thermal electric approaches, used to date by the
community, and develops some possibilities for future design considerations --
specifically, the experiences with sodium as a heat transfer fluid and its
limitations as an energy storage medium led to the suggestion of a sodium/
salt system.

The conclusions are:

1 - Thermal storage is a very important charateristic of solar
      thermal systems. This fact is often overlooked.
2 - Sodium is an excellent heat transfer fluid and has
      demonstrated its usefulness as a receiver coolant.
3 - Molten salt is a much better thermal storage medium than
      sodium.

IMPROVEMENTS IN MEASURING EQUIPMENT

Andreas Brinner, DFVLR

SUMMARY

The three years of operational experience with the measurement equipment of the SSPS-CRS has demonstrated that three main areas need improvement:

1. The measurement sensors should be calibrated periodically to guarantee the availability and accuracy of the measurements.

2. Additional measurement equipment should be installed to separate the measurement sensors from the control sensors and to allow measurement of the beam quality of all the heliostats, optical power into the cavity, spillage losses, and the stair-step of the circumsolar factor by hand.

3. The control and measuring systems should be separated completely, and control and data evaluation programs installed in their own computers. The data acquisition system should have its own front-end computer to receive the data evaluation of the measurements.

The appendix of the report contains the results of two calibration campaigns on the thermocouples and flowmeters. The examination of 12 relevant thermocouples resulted in deviations somewhat above the guaranteed tolerance of the NiCr-Ni sensors used. But with the help of the described polynominal and special coefficients for every thermocouple, these deviations can be corrected. The examination of the most important flowmeter, LK01CF01 of the sodium circuit, resulted in deviations in the range of 2.79% up to 11.29% less displayed. The deviation depends on the absolute value of the flow in m^3/hr.

The deviations stated are much higher than the expected inaccuracy of ± 2.5% over the whole measuring range. This flowmeter has been exchanged for a new, exact calibrated one.

POTENTIAL FOR IMPROVEMENTS

Ricardo Carmona and José Martín, ITET

SUMMARY

There are lessons to be drawn from the SSPS experience that may help
evaluate the potential of central receiver technologies. Our experience
confirms that sodium-cooled recievers can collect solar energy efficiently and reliably;.it also confirms the need for large scale , not only on
the basis of manpower considerations but also because of the relative
impact of plant loads and increased efficiency of larger subcomponents.

Discussions of possible improvements should start with an assessment of
what the performance of the CRS has been. 'Efficiencies' evaluated at
some instant are usually misleading. Operating days can be found when
noon efficiencies have been about 13%, even when the field was soiled and
the PCS was operating at half of its rated capacity. With a clean field,
and operating at rated high power, the efficiency could be as high as
15%.

A different picture appears when 'the calculations are made on the basis
of an entire operational day. On a day which was deemed 'good', the
overall efficiency has been measured as 8.1% with 25350 kWh potential energy offered to the heliostat field and 2065 kWh generated. These num-
bers are disappointing, particularly with the CRS in house plant load at
about 1472 kWh. Optimistic extrapolations are possible: the average PCS
efficiency for that date was only 17%. This weighed heavily on the overall generation and could certainly be improved.

Unfortunately, few days were 'good'. To discuss possible improvements,
it is helpful to differentiate between those which raise the average efficiency in a good day, and those which raise the plant availability.

With a modest effort on the pressent CRS only minor efficiency improvements are possible. A higher average field reflectance may be achieved
by either increased washings or antisoiling coatings, and spillage may be
reduced through corrections in tracking and better image quality. Major
efficiency improvements could be possible with a new PCS and, to a lesser
extent, with a redesigned field layout. If all of these changes were implemented, the overall daily gross efficiency could realistically be
raised to 18%, even for a plant of the size of the CRS.

74

Availability improvements are harder to quantify. A better field/receiver match and, of course, a suitable PCS would help raise availability. Other recommended measures are software modifications to add flexibility to normal operation, start-up and filling-up sequences, lightning protection, better heliostat images and tracking accuracy, and an improved signal for the feedforward control action for the receiver. Many of these measures can be accomplished realistically in the present CRS.

Operation of the receiver at higher than rated temperature may reduce or eliminate the losses which arise now because hot sodium must be cooled wastefully. However, future designs must take thermal inertia effects into account, so that plants can be operated effectively within the limits intended in the design.

It has been proven that sodium-cooled receivers can collect solar energy efficiently and reliably and that control technology can insure safe transient operation even for high flux (i.e., very efficient) receivers. It is not reasonable to expect more dramatic increases in the reciever efficiency. A plant which is about an order of magnitude higher than the CRS can operate at an overall gross efficiency of 20% on a good day. Extrapolation to even larger plants is not wise.

Energy storage in sodium is not an elegant or economical solution and the relative impact of plant loads must be lowered substantially. These disadvantages must be weighted against those of alternative receiver technologies. Finally, a major effort needs to be made in the matching of subcomponents and improving system reliability. Control technology can make a sound system work, but it cannot make it sound.

SYSTEM CONSIDERATIONS

Clifford S. Selvage, ITET

SUMMARY

In this paper the author has drawn upon the potpourri of evaluation
efforts of the SSPS-CRS system, most of which is presented in this CRS
report in an effort to place each of the evaluation conclusions in a
useful perspective.

What we see is a recognition of the useful attributes of the receiver
coolant - sodium - as well as its real (not imagined) short comings. The
single real advantage of a central receiver thermal system is its ability
to store energy in a thermal mass. Sodium is not the best material for
that purpose, but it is a superb receiver coolant.

Molten salt is an excellent storage material, but not the best heat
transfer fluid - hence the obvious suggested marriage.

VOLUME II:

DISTRIBUTED COLLECTOR SYSTEM

DISTRIBUTED COLLECTOR SYSTEM

The 500-kW$_e$ DCS plant consists of two types of line focus parabolic trough collectors, organized in three separate fields. Thermal energy is collected by the parabolic troughs using a high-temperature oil which is pumped to the top of the storage tanks, from which it can be pumped to a steam generator to produce steam for the DCS steam turbine. Low-temperature oil is returned from the steam generator to the bottom of a main storage vessel where it can then be pumped back to the collector field.

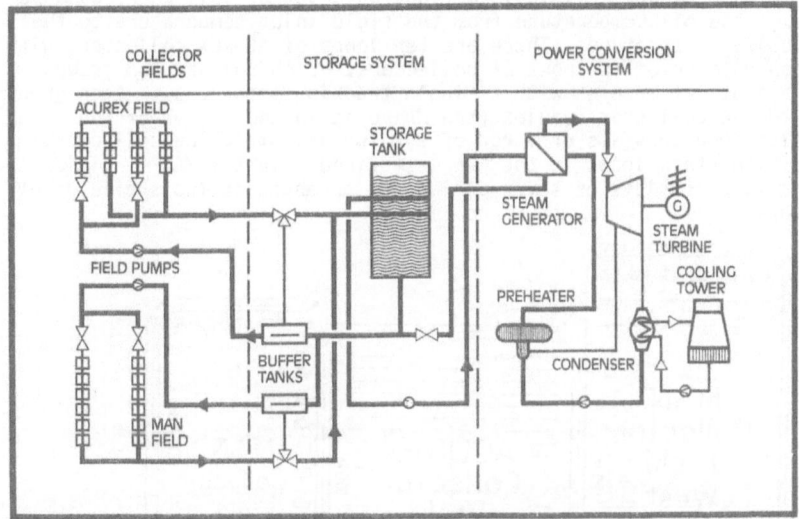

As shown, the DCS plant consists of three major systems: collector fields, storage system, and power conversion system. The main design data are given in Table I.

The middle field consists of one-axis tracking collectors covering 2674 m^2, oriented east-west, made by ACUREX (USA), Model 3001. The collector uses thin glass mirror (0.6-0.88 mm) with excellent optical performance.

The west field is equipped with MAN Helioman 3/32 model collector modules over 2688 m^2, two-axis tracking with 4-5 mm thick back-surface mirror hot-formed float-glass and a total area of 2688 m^2.

A third field of the same type of Helioman but improved, the east field has been incorporated at the plant in March 1984, increasing the DCS solar multiple to close to one by the added 2244 m^2 of collector area.

A loop in the collector field is defined as a set of collectors capable of raising the oil temperature from the field inlet temperature to the field outlet temperature. There are ten loops of ACUREX collector, with each loop made up of two rows of collectors. Each row has two groups (i.e., a group of modules with a single tracking motor), with each group having twelve collector modules (the drive is in the center of the group). A MAN west loop consists of a row of six two-axis-tracking collectors; there are fourteen loops in the MAN west field. In the MAN II field (east field) a loop consists of seven collectors arranged in two subfields of five loops.

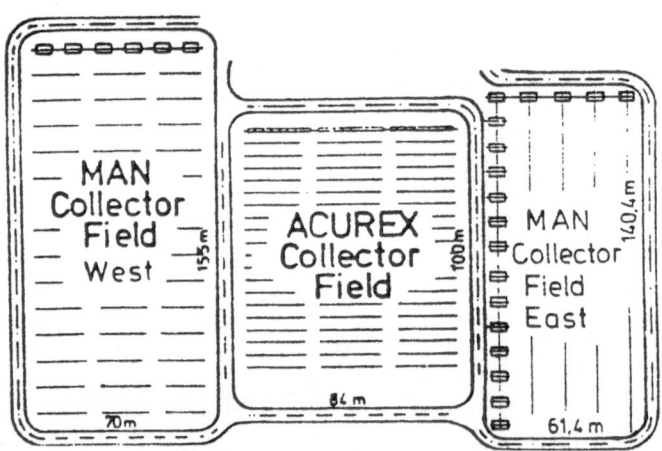

Solar radiation falling onto 7602 m^2 of the collector fields with short focal length is concentrated onto glass-covered, black-coated metal tubes containing thermal oil (Santotherm-tt) as a working medium to which the heat is transferred. Inlet oil temperature to the collector field is normally 225^{0}C and outlet oil temperature from the field is 295^{0}C.

Thermal energy storage for the DCS consists of
a single thermocline storage vessel. This
vessel is a vertical cylindrical shell with
dished headers welded to both ends. It stands
about 15 m high, has a diameter of about 4.2 m,
and a total volume of 176 m³. There are two
diffuser manifolds near the top of the vessel
and one diffuser manifold near the bottom. Hot
oil (295°C) enters the vessel from the collector
field through one of the upper manifolds and is
drawn from the vessel to supply the steam generator from the other upper manifold. Also, the
lower manifold is used to draw oil from the vessel to supply the collector field. The entire
vessel volume cannot be used for oil since a nitrogen ullage volume or
blanket is at the top of the vessel to prevent oxidation of the oil, control pressure in the vessel, and accommodate changes in the fluid level.
The actual working volume of hot oil for plant operations is 115 m³. The
design of this thermocline storage vessel is similar to the one used at
the U.S. Coolidge, Arizona solar-power irrigation facility, which began
operations in September 1979. There are two buffer tanks, one for each
of the ACUREX and MAN collector fields, which are considered as part of
the DCS storage system. These tanks are horizontal cylindrical shells
with dished ends that are about 2.5 m long and 1.0 m in diameter, each
with a volume of 1.5 m³. They are used to prevent cold oil from the collector fields from entering the thermocline storage vessel. Low temperature oil from the collector fields are independently bypassed around the
thermocline vessel, thoroughly mixed in the buffer tanks with other oil,
and recirculated through the fields until each field outlet temperature
is the required 295°C. Once each fields' outlet temperature reaches 295°C
the flow from that field can be supplied to the thermocline storage vessel.

The second stage, the Dual Medium Storage Tank (DMST) is of a dual medium
storage type, i.e., thermal energy is stored by means of thermal oil and
cast iron slabs. The tank consists mainly of a vertically standing steel
vessel, which contains a stack of 115 slabs of cast iron. The vessel is
built up in three parts, a lower, middle, and upper part, which are all
insulated and which are mounted on a base frame. From a height of + 3190
mm upwards, the 115 round and finned slabs of cast iron are piled up to a
height of 16646 mm. The cast iron slabs have a weight of 358 t and act as
the heat storage medium, whereas the thermal oil which circulates through
the slabs acts mainly as the heat transfer medium.

For loading the system, hot thermal oil will be pumped through the vessel and returns to the collector field. When passing the iron slabs from the center to the rim and back to the center, the oil transfers its energy to the iron.

For unloading of the system, cold thermal oil will be pumped through the vessel and returns to the consumer. When passing iron slabs, the oil extracts the stored energy.

The heat transfer and power conversion system has been developed with three heat transfer loops. The first loop extracts low-temperature ($225^{\circ}C$) from the bottom of the thermal storage tank, circulates it through the collector fields, and returns it at a temperature in the range of 275 - $295^{\circ}C$ to the top of the tank. In a second loop a boiler takes the hot oil from the storage tank, exchanges the thermal energy to the steam loop, and returns the cooled-down oil to the tank. The third loop circulates water through the PCS.

The major subsystems of the power conversion system are the steam generator and the steam turbine/generator unit. The steam generator consists of separate economizer, evaporator with a steam/water separation drum mounted above, and superheater. Feedwater enters the tube and shell economizer at $136^{\circ}C$. Heated water then flows to the evaporator, which produces a water/steam mixture which flows upward from the evaporator into the drum via tubes located in the inner core of the connecting pipes. Water from which the steam has been separated in the drum flows downward from the drum back to the evaporator via tubes located around the periphery of the connecting pipes. The saturated steam from the drum passes via water/steam separators to the tube and shell superheater. Hot thermal oil from the thermocline storage flows in series through the superheater, evaporator, and economizer returning to the thermocline storage or to the collector field. The heating surface for the economizer, evaporator, and superheater are 40 m^2, 131 m^2, and 42 m^2 respectively. All these vessels have 14 mm wall thickness, and the heat exchanger tubes outside diameter are 15 mm.

The steam turbine for the DCS plant is an eight-stage condensing turbine with one extraction for the deaerator. The turbine drives the electric generator through a reduction gear of the single reduction parallel shaft type. This gear is connected to the turbine by a tooth coupling and to the electric generator by a flexible rubber bush coupling. Selected design features of the steam turbine are:

inlet pressure	25 bar
extraction port pressure	3.4 bar
exhaust pressure	.07 bar
number of stages	8
turbine speed	9962 rpm
number of speed reductions	1
output shaft speed	1500 rpm
calculated efficiency of turbine and gear	23.9%

The electric generator is an air-cooled, 4-pole type. The poles of the rotor are mounted directly on the shaft, and the exciter unit is positioned outside the generator bearings. Selected design features of the electric generator are:

rating	713 KVA
output at .8 power factor	577 kW
efficiency at .8 power factor	95%
voltage/phase	400 V/3 phase
frequency	50 Hz

The DCS control and data acquisition subsystem is designed to demonstrate the DCS plant's operational behavior in interconnected grid or stand-alone modes. Except for some operations during start-up and shut-down, all operational modes can be controlled automatically.

The calculated overall efficiency of the turbine/generator unit is 22.7%.

DCS Main System Design Features

Collector Effective Area	5264 m^2
Collectors	14 loops with double-axis tracking MAN modules, "Helioman" 3/32
	10 loops of improved ACUREX single-axis tracking collectors with thin glass covered reflectors
	2 x 5 loop with double-axis tracking MAN module Helioman 3/32
Storage	Thermocline storage, equivalent to 0.8 MWh$_e$
	Dual medium storage tank, equivalent to 0.4 MWh$_{th}$
Heat Transfer Fluid	Thermo-oil for plant operation below 0°C and plant start-up at 50°C
Power Conversion	Adapted Stal-Laval TGC 8/2500 steam turbine, cycle efficiency 22.75% (calculated)
Safety Precautions	Uninterruptable power supply, lightning protection, provisions against earthquake damages
Design Lifetime	10 years
Guarantee	90% performance guarantee at design point at acceptance

Table I

HISTORICAL ASSESSMENT OF THE SSPS - DCS PLANT PERFORMANCE

INTRODUCTION

The following specific evaluation reports address the actual performance of DCS systems through the years 1981 - 1984. This performance is reviewed, analyzed and then summarized in the work by P. Wattiez, with the title of OPERATIONAL PERFORMANCE HISTORY. This work examines the fundamental questions confronting the solar plant design engineer when initiating the design of a new distributed collector system: -- how does this type of solar plant behave? what is the distribution of the energy? and, where does the energy go? All of this analysis is based on the recorded monthly data of 1981 thru 1984 and is presented in graphic form providing easy to follow understanding of the above questions.

Conclusions from Wattiez's work are:

1 - The overall system net efficiencies are lower than expected for all periods evaluated.

2 - The collectors thermal efficiencies are close to expectations.

3 - Thermal to electrical conversion efficiencies are much lower than expected.

4 - The addition of 70 two-axis tracking collectors in 1983 improved the overall system performance as a result of increasing the solar multiple (now close to 1).

The last of the above conclusions (4) is strengthened and expanded on in the evaluation by K. Schreitmuller, again using the recorded monthly data in the report INPUT/OUTPUT DIAGRAMS OF THE SOLAR FARM SYSTEMS. This work illustrates the operational behavior of specific subsystems and components under "normal" and "good" operational conditions, and leads to a second report in which the effect of the addition of the third collector field is examined and the improvements in that field installation which is a result of learning from analysis of performance in the original two-axis tracking collector field.

The conclusions of these two reports are:

1 - Input/output diagrams can help the understanding of a plant's behavior.

2 - More energy is available for collection with a two-axis tracking collector than a single axis collector (offered energy).

3 - The SSPS single axis collector was better able to utilize the energy available to it than the two-axis system.

4 - Considerable improvements were made in installation of the new two-axis collectors, reducing thermal losses.

The unique opportunity to compare two collector fields which use the same type of collector, with improvements in both the control system and the method of collecting the thermal energy, was accomplished by three members of the ITET: J. Sandgren, P. Wattiez, and M. Andersson, and is reported in A COMPARISON OF MAN-E AND MAN-W FIELD BEHAVIOR.

The conclusions of this evaluation are:

1 - The newer two axis collector field MAN-E has a thermal efficiency of approximatly 30%.

2 - The original two axis collector field MAN-W has a thermal efficiency of approximatly 21%.

3 - The improvement in efficiency with the newer system is offset by an increase in electrical consumption because of increased pumping power.

Through evaluation and analysis of the plant operation and performance it was possible to develop a system simulation program so that operational conditions can be simulated and the performance for those possible conditions can be observed. The program thus developed, the SOLAR ENERGY SYSTEM ANALYSIS MODEL or SESAM, has been used and is compared to actual observed performance. M. Andersson performed this work and prepared the report, SYSTEM SIMULATION.

The conclusion is:

1 - System simulation is an excellent tool for evaluating proposed hardware changes and/or proposed operational methods.

DCS - OPERATIONAL PERFORMANCES HISTORY

Pierre Wattiez

SUMMARY

In three years of operation with the weather conditions of the Almería
"Plataforma Solar", the SSPS-DCS plant has provided valuable performance
data and experience. Investigations of operation requirements and experiments with two different technologies, a solar system and a steam
power system, could be done.

This paper presents the major operation results from the evaluation of
plant performance data over the three operational periods. The first was
influenced by plant and subsystem acceptance tests, including a period of
functional adjustment as well as operational mode training, and permitted
the second year to start with the operational objective of maximization
of electrical production. The third year was then devoted to specific
test and evaluation programs dealing with the behavior of new components
such as collectors and storage.

First Year: October 1981 - September 1982

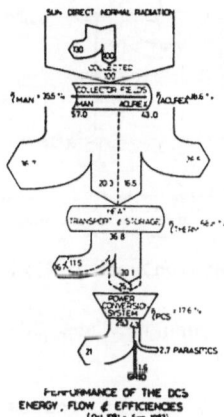

PERFORMANCE OF THE DCS
ENERGY, FLOW & EFFICIENCIES
(Oct 1981 - Sep 1982)

The acceptance phase and adjustment program marked the
first plant operation year. Performance data collected are only partially complete; some have been estimated. Evaluation of the plant performance shows an
electrical energy production of 178 MW$_e$h (gross) with
an annual gross plant efficiency (electrical energy
production/solar energy offered to the collector field
in track position) of 4.3%. The yearly collector
field efficiency, defined as the ratio between the
thermal energy produced divided by the solar energy
offered to the field in track position is 39 and 36%
for the single-axis and dual-axis tracking collector
fields.

Second Year: October 1982 - September 1983

Reaching the record of 234 MW$_e$h gross produced in this year, the DCS
plant achieved an annual gross efficiency of 4%. Weather conditions over
over the year provided 40% shining and clear days and 22% covered days.

87

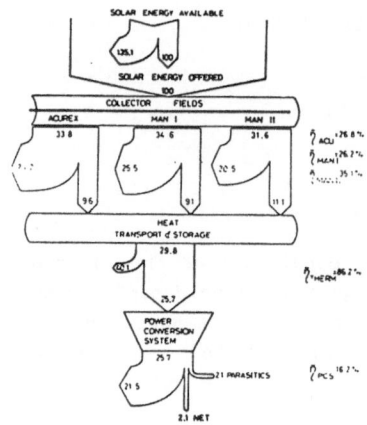

ENERGY FLOW CHART AND EFFICIENCIES OF DCS PLANT
FOR NET PRODUCTION PERIOD
(Mar.- Apr.- May -Jun, 1984)

With those conditions, the collector field performances annual efficiencies were 31 and 26% for the single-axis and dual-axis tracking collectors. The storage subsystem was credited with an annual efficiency of 83% (thermal energy in/thermal energy out).

Third Year: October 1983 - September 1984

The major event which occured during this last operational period was the integration of a third collector field which increased the plant solar multiple (ratio between the nominal field thermal power output and the nominal thermal power required to deliver rated power) close to one. A selected period was chosen (March - June 1984) considering the collector field availability and weather to estimate the annual gross plant efficiency at 4.2%. The collector field performance for this period showed 27, 26, and 35% for the single-axis, and dual-axis west and east collector fields. The storage subsystem presented a thermal efficiency of 88% during the year.

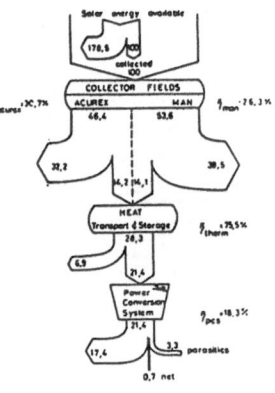

Energy flow chart and efficiencies
of DCS plant
for 12 months of operation.
(Oct. 82-Sep. 83)

CONCLUSION

Considering the operational conditions for each period, the evaluation of plant performance data reveals the energy losses during net electrical plant operation for the last two periods, and identifies a respectable plant thermal efficiency (thermal energy available at the storage outlet/solar energy offered to the collector field in track position) above 25%.

In conclusion, considering three years of operational results, the DCS daily efficiencies are as follows.

	%							
solar energy available	100							
solar energy 300 W/m²	94-95	100						
solar energy offered to the field in track position	70-75	75-79	100					
solar energy offered to field sending thermal energy to storage	54-56	57-59	75-77	100				
thermal energy produced by collector fields	20-22	21-23	29	37-39	100			
thermal energy stored	18-19	19-20	25-26	33-34	86-90	100		
thermal energy used for electric energy production	16-18	17-19	23-24	30-32	80-82	89-95	100	
gross electric production	3-4	3-4	4-5	6-7	15-18	17-21	19-22	
net electrical production	1-2	1-2	1-3	2-4	5-9	6-11	7-11	

From a review of the energy losses through each step of the solar thermal system, the possibility of a daily plant efficiency (gross/solar energy offered) of 10% appears achievable with equipment similar to that in the DCS plant.

88

INPUT/OUTPUT DIAGRAM OF THE SOLAR FARM SYSTEMS

Konrad R. Schreitmüller, DFVLR

SUMMARY

Solar thermal power plants, though very complex in their instantaneous behavior, show often rather straightforward overall performance indices. To determine those, the input/output diagrams of the total plant and the various subsystems have been investigated.   Only 'good' operational data. have been taken into account, thus avoiding the effects of breakdowns, tests, or other causes for bad operation. The results show clearly

- the higher mean daily offered energy on the two-axis tracking MAN collectors (Figure 1)
- the better utilization of the offered energy by the ACUREX collectors due to the lower losses and capacitances of the intermediate piping (Figure 2)
- the low net electric energy output with storage energies below approximately 5 MWh, thus recommending a modified operational strategy (Figure 3)

By means of these I/O diagrams an assessment of the overall behavior of the plant, whether in 'ideal' operation or with different climatic conditions, is easily possible.

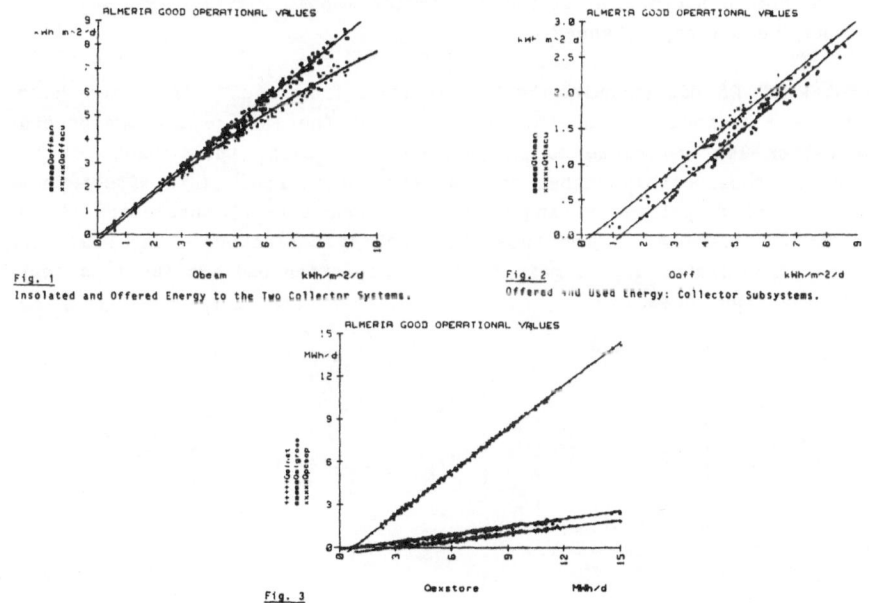

Fig. 1
Insolated and Offered Energy to the Two Collector Systems.

Fig. 2
Offered and Used Energy: Collector Subsystems.

Fig. 3
Storage Output and Used Energy: Power Conversion Subsystem.

89

EVALUATION OF THIRD FIELD PERFORMANCE

Konrad R. Schreitmüller, DFVLR

SUMMARY

Optical and thermal characteristics form the main features of a solar energy system. Thus, the accurate determination of these values is most important for the proper evaluation and understanding of the plant. This report deals with the results of the DFVLR measuring campaign in Fall 1984.

The DCS third field consists primarily of

- the forward piping of the North- and South-fields, respectivley.
- the ten loops with seven Helioman collectors each.
- the return pipe to the dual medium storage tank.

The following characteristics have been measured:

- the temperature-dependent thermal losses of each part of the piping.
- the flow distribution on the particular loops.
- the thermal capacitances.

A total of 26 distribution tests, lasting from one to fourteen hours each, were performed. The thermal losses of the main piping are according rather low and close to the theoretical value, while those of the loops are considerably higher and far more scattered. This effect indicates the difficulties arising with the proper thermal insulation of the complex construction of the two-axis tracking collectors. The remaining values (flow distribution, gain factors, capacitances) are far less scattered than the losses.

COMPARISONS OF MAN FIELDS' BEHAVIOR

Jonas Sandgren, Pierre Wattiez, and Mats Andersson, ITET

SUMMARY

The new MAN field installed at the SSPS site during 1984 is compared to
the original MAN field using data collected during April - June 1984.

The main differences between the fields are a layout that minimizes the
field pipe lengths, especially on the hot side, improved insulation, and
elimination of the long field pipe support (fins) of the old field. The
field control was also improved and the new field shows a more stable
operation characteristic than the old field.

The total energy of the direct irradiation during the day and total energy gained by the fields was selected as a basis for the evaluation. Efficiency here means the quotient thermal energy gained by the fields/
direct irradiated energy.

The analysis shows that the new field features considerably lower losses,
a marginally higher conversion efficiency and, as a result of this, is
capable of a considerably higher thermal energy production. The improvement is in the region of 10% on clear days and is higher on days with energy at low irradiance levels. Nevertheless, the new field consumes
twice as much electrical energy as the old field in producing 1 kWh of
thermal energy.

CONCLUSIONS

- The new MAN field (MAN-East) has operated with an efficiency of 30%.

- The corresponding figure for the old field (MAN-West) was 21%.

- The increase in thermal energy collected by the new field was deterred
  by the increase in internal electrical consumption.

SYSTEM SIMULATION

Mats Andersson, ITET

SUMMARY

Computer simulation programs can be extremely useful tools in the design
and analysis of physical systems. The effects of various parameter
changes on the performance of a given system can be determined rapidly
and at a relatively low cost. Several simulation models have been developed for the design and analysis of solar systems. One program that has
been used rather extensively is TRNSYS ('A Transient System Simulation
Program') which consists of a collection of subroutines that model subsystem components. This code is highly versatile but requires some expertise in programming.

The simulation code presented in this paper, SESAM ('Solar Energy System
Analysis Model' for distributed collector systems), is a detailed simulation model of the DCS plant in Almería, Spain. It differs in that it was
developed to model a specific solar plant that exists and is in operation. Some of the required features were a high speed of execution, easy
of use, and to be realisitic in its representation of the entire plant.

OPERATION STRATEGY

The effect that a change in operation strategy may have to the amount of
collected energy during a day is difficult to study as one day of operation cannot be repeated with exactly the same conditions. With a simulation model this is possible; three parameter changes are presented:

- Irradiance level for start-up
- Operation at different outlet temperatures
- Soiling and its effect on energy collected

CONCLUSION

A simulation code is an excellent tool for studying possible hardware improvements. By first testing the effect in simulation, money and work
can be saved. Two different improvments are studied in this paper:

- New strategy for controlling the collector field bypass valves
- Reduction of thermal inertia in the two-axis tracking collector field
  (MAN)

SURVEY OF PLANT LOSSES

INTRODUCTION

The understanding of a plant's behavior under actual operational conditions requires knowledge of the plant's losses. This seems particularly true with a solar thermal distributed system because of the distributed optical subsystems, the extensive piping required to interconnect the individual collectors, the need for a thermal storage subsystem, and the thermal to electrical conversion subsystem.

The following collection of specific evaluations addresses each of these identified loss areas with some helpful theoretical analysis, and considerable analysis of the collected operational performance data. Conclusions of each evaluation are provided, summarized in this introduction, contained in more detail in each of the reports, then reviewed and used in the evaluation report POTENTIAL FOR IMPROVEMENTS in Section 7 of this volume.

Starting with a theoretical discussion of the optical subsystem, with the objective of identifying the potential loss areas, J. Martin and R. Carmona move into a practical analysis of the specific optical characterisitics of the two parabolic trough collectors in the DCS fields. Finally, using the results of extensive measurements, they report on each of these loss areas and develop conclusions which are helpful in effecting improvements with design of new collectors, and are certainly useful in the evaluation report POTENTIAL FOR IMPROVEMENTS.

The conclusions of this evaluation are:

1 - Alignment and tracking errors have developed in the single axis tracking collector system which seems to have caused an 8% reduction in thermal efficiency.

2 - Cosine and end losses penalize a one axis tracking system.

3 - The Black Chrome absorptive coatings on the single axis tracking collectors have degraded. The Solartex absorptive coating used on the two axis tracking collectors have not degraded.

Proceeding with the energy transfer process; that is, from optics and energy absorption into the thermal energy collection and transport, the report THERMAL LOSSES OF THE COLLECTOR FIELDS by H. Jacobs and R. Carmona discusses the extensive measurements conducted during the three years of test and evaluation, again contributing to possible improvements with future designs.

Conclusions are:

1 - Thermal losses for the SSPS two axis tracking field are
   approximatly 620 kw and 350 kw for the single axis field.

2 - Thermal losses in the passive piping of the two axis field
   account for 12.7% of all available energy. Losses in the
   passive piping of the single axis field are 0.8% of all
   available energy.

3 - It should be possible to design a two axis tracking field
   with much reduced losses that could be cost effective.

Thermal energy storage fills an important function in the distributed
collector type system, as all of the collected thermal energy is trans-
ported from the collector fields to the thermal storage subsystem, then
to the thermal to electrical conversion subsystem. Consequently, all
the thermal energy goes through the thermal storage subsystem. The
thermal behavior of the SSPS thermocline storage subsystem was analyzed
and reported on in the report DCS THERMAL STRATIFICATION. The evaluation
was accomplished by M. Andersson.

The conclusions are:

1 - The thermocline stratification storage worked well. The
   thermocline was easly established and was maintained over
   long periods.

2 - Approximately 4% of the energy sent to storage is lost, with
   the major loss element insufficient insulation.

The last subsystem in the SSPS distributed collector system is the power
conversion subsystem - the thermal to electric conversion subsystem. This
subsystem consists of an oil to water heat exchanger or steam generator, a
steam turbine, and the AC generator. The report of the evaluation of the
PCS accomplished by R. Carmona and H. Jacobs is called LOSSES AND PERFORM-
ANCE OF THE POWER CONVERSION SYSTEM.

The conclusions are:

1 - The PCS losses are primarly Carnot with most of the energy
   dissipated by the cooling tower.

2 - Other thermal losses could be reduced up to 10% by improvement of the insulation.

3 - The efficiency of the PCS is less than 20%.

To complete the survey of losses of this distributed system, the problem of plant internal electrical consumption (parasitics) must be evaluted and reported. P. Wattiez accomplished this completing the work, DCS INTERNAL ELECTRICAL CONSUMPTION.

The conclusion is:

1 - The two axis tracking collectors consume 12% more electrical energy per KWh th collected than the single axis tracking collectors.

All of the above summarized evaluations provided indications of how improvements could be accomplished. This provided a major input to the POTENTIAL FOR IMPROVEMENTS evaluation.

OPTICAL LOSSES

José G. Martín and Ricardo Carmona, ITET

SUMMARY

The DCS optical losses may be profitably discussed in terms of a four-factor formula: the optical efficiency is taken as the product of reflectance, intercept factor, transmittance, and absorptance. This paper reports on the results of measurements or estimates of these factors, and discusses the implications for operation and washing frequency.

Systematic reflectivity measurements reported elsewhere in this review have been complemented by detailed measurements in one of the ACUREX loops: the reflectance is slightly higher at the top facets than at the bottom, and at the middle module than at those close to the edges of the field.

The intercept is affected by sunshape, tracking, and displacement errors. From sunshape measurements made on site during the last two years, it has been estimated that the effect of circumsolar radiation on the intercept is less than 1%. The effect of alignment and tracking errors is more important: these may account for an 8% reduction in thermal efficiency of the ACUREX field.

The ACUREX coatings have degraded and a value of 0.8 may serve to characterize the field absorptance. The MAN Solartex coatings show no deterioration, although careful measurements indicate a small absorptance loss in the direction facing the mirrors.

Soiling affects transmittance to a lesser degree than it affects reflectance.

The field thermal efficiency drops faster than the reflectance. It drops faster for days when the irradiance is low. It also drops faster for the MAN field, because of the higher thermal losses in this field. It is possible to estimate what the minimum irradiation must be to deliver energy to the tank at a given temperature and flow rate, and to quantify the effect of increasing the frequency of the washings.

THERMAL LOSSES OF THE COLLECTOR FIELDS

Heinz Jacobs and Ricardo Carmona, ITET

SUMMARY

To improve the operational efficiency of the DCS plant, it is necessary to determine the sources of the thermal energy losses. The study described in this paper is directed at the energy loss that occurs when thermal oil is pumped through a desteered system. The tests and theoretical analysis performed were directed specifically at the absorber tubes, pipework, and supports.

Cosine and end losses penalize a one-axis tracking system in comparison with a two-axis one. In the case of ACUREX, these losses account for (21%) of the available energy. For the MAN, these losses are negligible during operation.

At noon, the MAN collectors have a slightly higher optical efficiency (68%) than ACUREX (65%). In terms of the daily losses and aside from shading and cosine effects, MAN daily optical losses (22% of the available energy*) are greater than those of ACUREX (19%) simply because the MAN field operates longer.

For normal operation (T_i = 215°C, T_o = 290°C), the total field thermal power losses are 620 kW for the MAN field and 350 kW for the ACUREX field. In terms of the total solar energy offered, the MAN losses amount to 23%, while ACUREX losses are 11%.

The MAN collector on the average collect 50% more solar thermal energy than the ACUREX collector. In spite of this, the total thermal energy collected by the MAN field is only 16% higher than the ACUREX field. The reason for the large difference between these results is, of course, the large thermal losses in the passive pipes in the MAN field.

The MAN passive thermal losses account for 12.7% of all the available energy; the ACUREX losses account for 0.8% of the available energy. It should be possible, and cost-effective, to design a field similar to MAN with less thermal loses.

---

* Available energy refers to the product of the pyrheliometer reading times the collector area integrated between sunrise and sunset.

THE DCS THERMAL STRATIFICATION TANK

Mats Andersson, ITET

SUMMARY

Thermal stratification (thermocline) tanks allow storage of energy at different temperatures by taking advantage of the thermal characteristics of a fluid. The benefits of a stratification tank can be shown by its use in the DCS plant:

-Using only one tank, two different temperatures are available, feeding the fields with low-temperature oil from the lower part of the tank and, at the same time, loading the upper part with hot oil for later use in the Power Conversion System (PCS).

-The thermocline (i.e., the zone where the temperature profile changes rapidly) is stable enough to withstand the dynamic operation during charging and discharging of the tank. Maximum electric energy can be produced by allowing the PCS to work at high temperatures until the tank is empty (i.e., the entire tank content is at 215°C).

The behavior of the tank during nonoperation is studied and discussed in this paper. The thermal losses are calculated to be between 15 - 20 kW, depending on the temperature in the tank. The stability of the thermocline is presented by showing the temperature profile inside the tank.

During mixing of the tank by pumping cold oil through the collector fields, energy is lost at a rate of about 800 kW.

CONCLUSIONS

By studying the energy flows into and out of the tank during operation days, the thermal loss is calculated to be approximately 4% of the energy sent to the storage tank.

LOSSES AND PERFORMANCE OF THE POWER CONVERSION SYSTEM

Ricardo Carmona and Heinz Jacobs, ITET

SUMMARY

This paper presents results of evaluation performed on the power conversion system (PCS) of the IEA/SSPS distributed collector system (DCS). The steam generator of the PCS is fed by thermal oil coming from the storage tank. The steam runs a turbine-generator system (Stal-Laval) with a nominal electric output of 577 kW.

Data from operational days are used to determine the energy need for start-up, which depends mainly on the time elapsed since the last operation. This is between 200 and 1200 kWh_{th}.

The losses of the PCS are also evaluated.

The losses during alternator operation consist mainly of the power to the cooling tower and thermal losses of the components. Both figures depend on the load.

One conclusion is that the energy loss during start-up reduces the total daily efficiency by more than 10%. A longer daily operational time, which could be a result of a higher solar multiple, could reduce this figure to less than 3%.

The power to the cooling system (70%) cannot easily be reduced because it is based on the Carnot process.

The thermal losses (10%) can be reduced by changing the hardware (i.e., better insulation of the components).

The actual efficiency, which depends on the load, is evaluated to be less than 20%.

DCS INTERNAL ELECTRICAL CONSUMPTION

Pierre Wattiez

SUMMARY

The operation of the DCS requires an amount of electrical energy which should be deducted from the plant electrical production. The energy delivered to the local grid represents the net plant electrical production.

The SSPS project plant contains two solar power plants with common building, services, and equipment causing some difficulty in assigning each plant its correct part of the common electrical consumption. There is some necessity of sharing certain electrical loads; for example, air conditioning in the common control room. Other loads have no relation to the proper function of these solar plants (i.e., the office building consumption).

This paper addresses the recorded SSPS common electrical consumption and distributes it as 61 and 39% for the CRS and the DCS plants. Further, it defines the correct DCS internal electrical load during activated and nonactivated plant periods.

Based on plant electrical loads measured by five instruments over the last two years, evaluation of the DCS subsystem electrical consumption is reported and defined. Their contribution to the plant load is as follows:

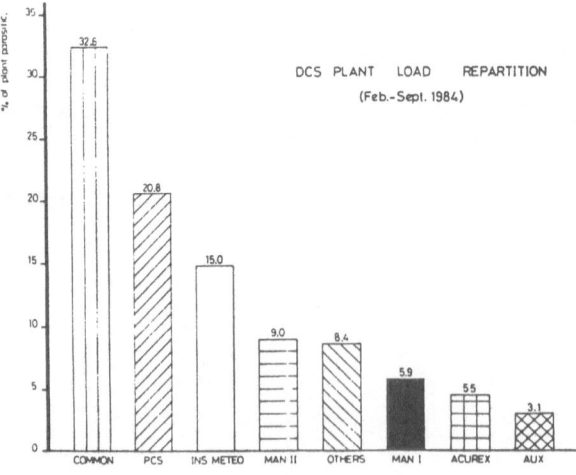

The integration of an additional 2240 m^2 of collector area has modified the plant internal electrical consumption. Analysis of the required electrical consumption per kWh thermal produced normalized to one square meter of collector area confirms that there is an advantage of the single-axis over the two-axis tracking collector (11.5 to13.0 Wh$_e$/m^2 for 1 kWh$_{th}$ produced/m^2).

The new dual-axis tracking collector field is evaluated in for its contribution to the increasing plant thermal energy produciton in comparison to its observed plant electrical load.

CONCLUSION

The design of a collector field should consider the optimization of the thermal energy to be produced in relation to its operational electrical energy. Considering the impact of the internal electrical load on the plant net efficiency, the concept of future DCS plants should avoid an excessive electrical consumption. Attempts should be made to make the solar system autonomous from the point of view of internal electrical consumption.

Any extrapolation of the SSPS-DCS plant load for future solar distributed collector system designs should be made with caution because of the large internal electrical consumption, as a result of the research nature of the SSPS.

POSSIBILITY OF AUTOMATIC CONTROL

INTRODUCTION

Experience with the SSPS distributed collector system over the three years of operation, tests and evaluation, has shown that automatic control of this type of system is essential in order to achieve a cost effective installation. Experiments using the SSPS DCS fields and a small on-site computer, have demonstrated that this type of control is achievable with a suitable control algorithm and a reliable collector array.

The two specific evaluation efforts reported in this section describe the experimental effort accomplished. The first report by R. Carmona, TEMPERATURE REGULATION, commences with a theoretical treatment of the control possibilities then provides details on improvements that have resulted and reports of lessons learned.

The conclusions are:

    1 - Automatic control of a DCS is practical and demonstrated.

    2 - Simple models for field response have been developed and used.

The second evaluation report contained in this section is, ADAPTIVE CONTROL OF THE ONE AXIS TRACKING COLLECTOR FIELD by E. Camacho, F. Rubio and R. Carmona, which reports on the progress made toward implementation of adaptive controls to the one axis tracking collector field and some experimental results.

The conclusion is:

    1 - An adaptive system was developed and it worked well.

TEMPERATURE REGULATION

Ricardo Carmona, ITET

SUMMARY

The performance of distributed collector systems is very dependent on the flow controls. Changes of insolation, inlet oil temperature, etc., require that the flow rate changes in such a way as to maintain the desirable output. If the changes involve large amplitude oscillations, the field performance suffers. Not only are thermal and pumping losses increased, but the relatively narrow band between the design outlet temperature and the temperature which triggers the alarm signal may be bridged by large amplitiude oscillations.

The first section of this paper describes the response of one of the fields (MAN) during an operational day in the summer of 1982 when the field oscillations are clearly exagerated. To make the field less 'jittery', it seemed desirable to make changes in the control system. These changes affected the setting of the minimum flow, and the governing control equation, as well as changes in the determination of the analog controller settings, interpretation of the irradiance of the feedforward action, and the elimination of the loop valves.

With a good analytical model, it is possible to optimize the field behavior. To develop such a model, a digital computer has been connected to the DCS. This paper presents a section describing the hardware changes implemented to superimpose special signals on the field.

As a side effect of the hardware implementation, the digital control of the fields has turned into a software problem. Details about the software and data about the field response during start-up and under simulated disturbances are presented.

Simple preliminary models for the field response have been developed in terms of lumped and distributed parameters, which are obtained using the so-called Powell algorithms. These are described and the predictions from the model are compared with actual field data. This development is also the basis for the consideration of adaptive controls. A report on the work performed is included.

A new field of two-axis tracking collectors has been installed at the site. Its design and controls take advantage of some of the lessons learned and are briefly described in this paper and some preliminary results are presented.

ADAPTIVE CONTROL OF THE ONE-AXIS TRACKING COLLECTOR FIELD

Francisco R. Rubio and Eduardo F. Camacho, University of Seville,
and Ricardo Carmona, ITET

SUMMARY

This paper presents an application of an adaptive regulator to control
the outlet temperature of the ACUREX distributed collector field of the
SSPS plant in Tabernas. The algorithm uses a recursive least square
identifier with a variable forgetting factor and a PI regulator. The
regulator uses a supervision level to improve the control.

The objective of the control system in the distributed collector field is
to maintain the outlet oil temperature at a desired level in spite of
disturbances such as changes in the solar, irradiance level (caused by
clouds or the time of day), mirror reflectivity, and inlet oil temperature.

The distributed collector field is a nonlinear system which can be ap-
proximated by a linear system when considering small disturbances. In
the design of any regulator the operating point of the plant must be born
in mind. However in a solar energy plant the operating point varies according to the time of day or disturbances caused by clouds, and therefore it is not possible to design a fixed regulator fully guaranteed to
work. Because of this a self-tuning regulator seems to be a good solution to this problem.

The plant to be controlled is described in Section 2 of this article.
Section 3 gives a description of the control algorithm used, and Section
4 comments upon results obtained.

RELIABILITY-AVAILABILITY AND MAINTENANCE

INTRODUCTION

A major question regarding solar thermal electric systems is the reliability of these systems and the quantity of maintenance required to keep them operating. Relatively complete maintenance records have been kept at the SSPS project, therefore providing the opportunity to evaluate the maintenance history, particularly of the DCS, and also to assess the reliability of the system and the subsystems. In conjunction with the records of the system and subsystem operation, which were kept by the Sevillana operators, and with the meteorological data, the true availability of the systems could be, and was, evaluated.

F. Ruiz, head of the Plant Operating Authority (POA), provided the much needed record and evaluation of the DCS, which is recorded in the report, DCS OPERATIONAL EXPERIENCE AND MAINTENANCE. This evaluation was performed by A. Cuadrado and C. Lopez, both of Mr Ruiz's staff.

The conclusions of this evaluation are:

1 - The two axis collector system is difficult to maintain and required much maintenance.

2 - The single axis collector system requires little maintenance.

3 - The electronics in the single axis system are poorly protected.

4 - There is insufficient automation in the total DCS.

5 - The auxiliary systems, fire-fighting and water treatment subsystems, have degraded badly.

6 - A larger than normal supply of replacement parts is necessary to maintain this type system in this type of remote area.

B. W. Swanson, a graduate student from the University of Arizona, performed a detailed analysis of the data collected on the DCS, collated this data, compared it to the meteorological data, and was then able to determine availability and reliability of this system. In the report of this evaluation, Solar Availability, Equipment Availability, Usage Factor, and Daily Usage Factor, are defined and then tabulated for each of the two types of collector systems. The conclusions confirm the observations of the previous report with a rigorous analytical approach.

The conclusions are:

1 - The single axis tracking collector system has high equipment availability and requires little maintenance. However, it cannot operate at low insolation levels.

2 - The two axis tracking collector system has lower equipment availability but can operate at low insolation levels. However, the longer operation capability could not compensate for frequent partial outages.

Maintenance records are necessary in performance of this type of analysis and are frequently incomplete or inadequate. More attention should be given to collection of this type of information.

The United States solar thermal program has sponsored the development of several distributed collector systems throughout the USA. Sandia National Laboratories, as the technical director of this program, was requested to add some comparative results from their evaluations. This was accomplished by E. C. Boes in the report, COLLECTOR FIELD MAINTENANCE; DISTRIBUTED SOLAR THERMAL SYSTEMS.

The conclusions presented by Mr. Boes are:

1 - Collector fields can and should be operated automatically.

2 - Maintenance costs can be reduced to about $1.00/m2 of collector area.

3 - Annual electrical consumption is 5 to 10 KW/m2.

A specific developmental equipment failure occurred with the single axis collector system that deserves discussion and a specific report. This problem area is related to the extra thin, tempered mirror, which is bonded to a steel sheet and subsequently bent into the parabolic shape required, when it is mounted into the structure. The problem was a debonding with the early produced mirrors, analysis of the failure and correction of the problem by the manufacturer. A detailed discussion of the stresses on the mirror/metal interface is contained in the report and a description of the problem solution is provided.

The conclusion is that the manufacturer can produce the extra thin mirrors and when applied in the manner as done with the SSPS, they can survive for long periods.

DCS OPERATIONAL AND MAINTENANCE EXPERIENCES

Antonio Cuadrado and Carlos Lopez, Sevillana

SUMMARY

This paper summarizes the experiences of the Plant Operating Authority with the Distributed Collector System of the SSPS plant.

After a short description of the system operation, comments are presented on the operational ability of the various susbsystems.

The high thermal inertia of the collector fields makes their operation difficult, especially on partly cloudy days. The field control is not the optimum solution for a system with long residence time and load dependent characteristics. The storage system is easy to operate. A desired feature would be the possibility to mix the contents of Tank I without passing the oil through the fields. The power conversion system and electrical system have a common characteristic -- insufficient automation. Remote monitoring and control have scarcely been applied. The data acquisition system shows a lack of flexibility to modify the format of data presentation and a rather long data refresh time (15 seconds).

The maintenance characteristics of the subsystems are presented next. The MAN collector fields take the main part of the maintenance effort put on the DCS. This is mainly due to design errors and bad maintainability of the modules. The ACUREX field and storage system have required little maintenance. The main problems with the power conversion system have been due to steam generator leaks and repeated contamination of the lubrication oil in the feedwater pump. The electrical system reveals poor workmanship and is badly documented. The data acquisition system has required normal computer equipment maintenance. In the auxiliary system, mention must be made of the degraded and/or broken sensors in the fire fighting station and insufficient capacity of the water treatment plant to supply the needs of the system.

MAINTENANCE, RELIABILITY, AVAILABILITY

Belinda Wong Swanson and Rocco Fazzolare, University of Arizona

SUMMARY

Availability of a device or system is generally used as a measure to judge its effective operation. However, there are many factors that cause nonoperation of solar systems, such as low insolation level and high wind velocity, in addition to equipment failure. Therefore it is desirable to find the usage factor, defined as the ratio of the collector field working hours to the possible hours of operable radiation level, and corrected by the equipment availability to account for partial operation of the collector field.

This study uses the operation and maintenance records of the ACUREX single-axis tracking and original MAN dual-axis tracking collector field subsystems for the first eight months of 1984 to find the usage factor of the systems, the solar availability, as well as the equipment availability. The most frequent repairs, causes of failure, and impact on system shutdown are identified.

Results from the study showed that the single-axis tracking collector system is limited to peak sun-hour operation. However its simple design has less components, therefore it has good reliability and requires minimal maintenance. On the other hand, the dual-axis tracking collector system has longer operation hours because it could work in the early mornings and late afternoons. Its design is much more complex, requires many components and controls thus increasing the chances of malfunction. It is less reliable than the single-axis system, it seldom operates at 100% capability, requires frequent maintenance and many labor-hours. The frequent partial outages could not be overcome by the longer operation hours, therefore its energy output capacity is generally lower than the single-axis system. In general it could be concluded that the single-axis system operates more effectively than the dual-axis system.

COLLECTOR FIELD MAINTENANCE: DISTRIBUTED SOLAR THERMAL SYSTEMS

E.C. Boes, E.C. Cameron, and E.L. Harley, Sandia National Laboratories

SUMMARY

This paper reports on recent operation and maintenance experiences with distributed solar thermal systems. Although some information on system-level operation and maintenance requirements will be included, the emphasis will be on reporting collector field maintenance experiences.

Operation and maintenance data has been compiled from three different distributed solar thermal projects for this presentation, including a total of 11 different collector fields. The three projects are:

* Coolidge Solar Irrigation Project
* Solar Industrial Process Heat (IPH) Project
* Modular Industrial Solar Retrofit (MISR) Project

All 11 of the systems to be discussed are located in the USA. For most of these systems, performance and maintenance data is available for at least a year. With the exception of the first project listed, the collectors in these systems represent "second-generation" technology.

The collector maintenance experiences for these systems have been quite variable. Annual collector maintenance costs have ranged from about $1 to almost $15 per square meter. The major factors causing this large range are the maintenance approaches adopted by the different system operators and different labor rates. Some operators have devoted a considerable amount of labor to "preventive" maintenance, while others have only provided a basic level of routine maintenance and have concentrated primarily on corrective maintenance as necessary for system performance.

The major conclusions of this presentation are these:

- Collector fields can be operated completely automatically; this has been demonstrated in several systems.
- Collector field maintenance costs can probably be lowered to an annual level of $1 per square meter of collectors at least for the initial years of operation.
- What effect longer periods of operation, such as 10 to 20 years, will have on collector field maintenance costs is not yet known.
- Annual electrical power consumption requirements are about 5 to 10 kWh per square meter of collector area.

MIRROR DELAMINATION

J. W. JACOB, G./ MERTENS, J. DECLERK

Glaverbel, Belgium

A mirror panel made with an extra-thin, tempered mirror, bonded to a steel sheet by means of an acrylic glue, has been developed by Glaverbel. This was accomplished by bringing together several high technology processes such as security glass for automobile windshields, high quality mirroring and chemical tempering. This was acomplished by Glaverbel in cooperation with the Acurex company and installed on the DCS collectors supplied by Acurex.

Following several years of operation, delamination of the glass/mirror layer from the steel backing was observed, causing an extensive investigation in an attempt to determine the cause and to developed corrective procedures in the manufacturing process to prevent reoccurance of this problem.

This report presents a detailed discussion of the stresses on the mirror/metal interface and a description of the solution developed. The conclusion in this report is that the manufacturer now can produce the extra thin mirrors and when applied in the manner as done with the SSPS/DCS they can survive for long periods of time.

POTENTIAL FOR IMPROVEMENTS

One of the more important objectives of the SSPS project was to determine from the experience with the plants what could be done to improve performance in designing another distributed collector plant. Much of the defined and performed tests, followed by detailed evaluation of the resulting data, was used in responding to the TOAB request for this evaluation report.

In order to determine what can be done to improve a plant, the deficiencies of the plant must be known and understood. This was accomplished with the SSPS DCS with most of the evaluation reported in Section 3 - HISTORICAL ASSESSMENT OF THE SSPS-DCS PLANT PERFORMANCE and in Section 4 - SURVEY OF PLANT LOSSES. The graphical presentations from Section 3 identify the areas for potential improvement and lead to concentration on those loss elements where there is some physical reason to believe improvement can be made.

Having measured and evaluated the losses, a team of ITET engineers, university professors and industrial designers was assembled, with the objective of examining each of the deficiencies for improvement potential. The results of this effort is summarized and reported by P. Wattiez, P. Toggweiler and M. Andersson in the report IMPACT OF DCS IMPROVEMENTS.

A summary of the conclusions is:

1 - Some improvement can be made with each of the steps of collection, transfer and conversion.

2 - The most significant improvement is with the PCS, where an efficiency improvement to an efficiency of 25% would result in a global net plant efficiency of 11%.

3 - With improvements made in most of the elements a suggested global net plant efficiency of 14% is possible.

Drawing again from the experience with the United States' Solar Thermal Program, E. C. Boes provides some reference possibilities in the report COLLECTOR FIELDS; POTENTIAL FOR IMPROVEMENTS.

The conclusion is that very important technological advances will result with continued development of DCS.

POTENTIAL OF IMPROVEMENTS

Pierre Wattiez, Peter Toggweiler, and Mats Andersson, ITET

SUMMARY

Results and lessons learned from tests and operation programs of the SSPS-DCS experimental plant have brought forward some areas of potential performance improvements. Illustrated by a daily energy stair-step, the behavior of major subsystems and their respective losses are evaluated and a series of potential possibilities for improvement are discussed and the potential impact on subsystem and plant performances are examined.

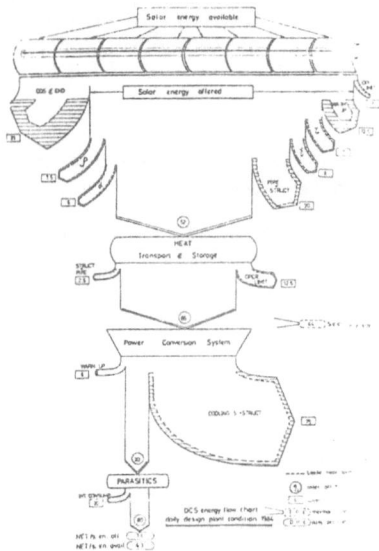

From the experiences gained, future design specifications are considered. One conclusion is that the design specifications in the future should be based on daily or annual performance characteristics.

The impact of collector area increase of 42% by the integration of a third field, caused the daily global plant efficiency to rise from 5.1 to 6.0%.

In order to judge the capacity of the plant using the three different collector fields, results are presented on the expected DCS daily global net efficiency using a single collector field.

From theoretical considerations of the upper limits of efficiency with a Carnot-type power plant, the technical and suitable limits of distributed collector power plant performances is made.

In the near term, with a reasonable improvement of the power conversion system to an internal efficiency of 25%, a global net efficiency of 11% is achievable. A value of 14% can be expected with additional changes of plant operational procedures.

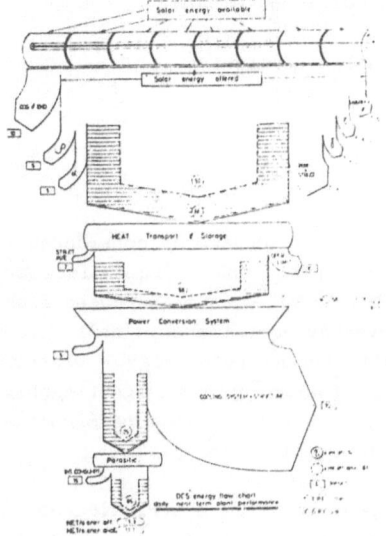

In conclusion, when comparing the measured and achievable performances of other types of renewable energy technologies, we see that the DCS-type plant can, in the near term, present some good reasons for applications of this technology, which would be competitive with other technologies.

115

COLLECTOR FIELDS: POTENTIAL FOR IMPROVEMENTS

Eldon C. Boes, Sandia National Laboratories

SUMMARY

Distributed solar thermal collector technology has undergone considerable
development during the past decade. For example, peak efficiencies for
parabolic trough collectors operating at 250 to 300°C have increased from
about 50% ten years ago to 70% today. Nevertheless, there is still considerable room for improvements to parabolic trough technology. Distributed point-focus solar thermal collector technology is significantly
less mature than parabolic trough technology, so relatively more significant technological advances are likely to be realized for it.

This paper discusses the potential for improvements of distributed solar
thermal collector technologies. For line-focus collectors, developmental
activities that can improve performance, durability and reliability, and
costs are separately discussed. In the case of point-focus technology, a
status report is given, and a brief discussion of this technology's most
important advantages relative to other solar energy technologies is
given.

Most of the information in this paper is presented in a concise, tabular
format to ease its assimilation.

The primary coclusion of this paper is this:

    Very important technological advances will result from continued
    development of distributed solar thermal collectors.

* This work is supported by the Division of Solar Thermal Technology,
U.S. Department of Energy.

VOLUME III:

SITE - METEO - ENVIRONMENTAL - AND SOILING CONDITIONS

SITE DESCRIPTION

The SSPS plants were constructed and are operated, on the Spanish Plataforma Solar near Almeria, in southern Spain. The specific location is: longitude 2°23'W and latitude 37°06'N. and the elevation is 500m. The site is in a semi arid area where agricultural efforts are frequently attempted, but have a rather poor success history. At the time this site was selected by the Spanish authorities the province of Almeria was a rather economically depressed area, having incurred three to five years of below normal rain fall and a slow industrial growth.

AREA OF THE PLATAFORMA SOLAR - 1980

Clearly, the Almeria area is located in the region of an expected high annual sunshine hours (2950h/y), is an international shipping port and has an international airport capable of receiving large cargo aircraft. All of these charateristics contributed to selection of this site and in addition the fact that weather data for the preceeding years was available from the Almeria airport.

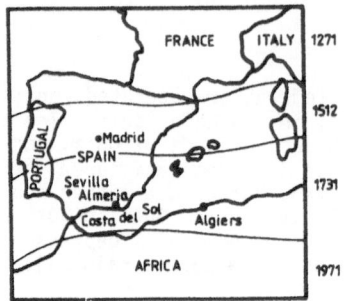

LOCATION OF SSPS SITE

Although the site is located in a valley some 40km from the Almeria airport, with a mountain range in between, it was felt that this weather data could be used as the basis for the solar system design specification. Unfortunatly the weather at the Tabernas site is very different than the weather at the reference airport.

The basic terrian in the area around the site effects the performance of the solar concentration systems by modification of the atmosphere with effluents from the area. There are industrial operations both to the east and the west of the site where significant particulate matter is emmited into the atmosphere frequently. In addition, much of the area around is very dry and sandy so that when the wind is strong particulate matter is carried into the air.

SURROUNDING AREA

Site security is provided by a guard force and the entire plataforma is fenced with a reasonably secure fence and gate system. Electricity is provided by the local electrical utility Compania Sevillana de Electricidad and telephone service by the local Spanish authority.

METEOROLOGICAL CONDITIONS

INTRODUCTION

Both the central receiver system and the distributed collector system of the SSPS project have been affected and conditioned by the site location during the three year operation and evaluation period. As with all solar systems, the weather conditions, including the solar insolation (irradiance), wind, rain and all other atmospheric conditions, are the key elements the systems performance. The system can only respond to these weather elements with the total energy produced some fraction of what is available. L. Castillo and M. Andersson have analysed all of the available insolation, wind, cloud presense and atmospheric transparency data collected over the operational time period and produced the report THE SSPS METEOROLOGICAL CONDITIONS 1982 - 1984.

The conclusions are;

1 - 1982 was a poor solar year.
2 - In the first six months of 1984 a total horizontal surface global irradiation of 900 kWh/m^2, which is what is expacted for this area, was received.
3 - In the winter months, fewer clouds are present after noon than before noon. The reverse is true in the summer.
4 - There are east winds every month but not west winds every month at the SSPS site.
5 - When there are west winds they propably exceed the operational limits of the plant.

SSPS METEOROLOGICAL CONDITIONS, 1982 - 1984

Lorenzo Castillo Garcia and Mats Andersson, ITET

SUMMARY

Insolation data and information concerning clouds and the transparency of
the atmosphere are important for solar energy applications. It is also
important to know the wind direction and force at a plant location due to
the influence that wind may have on the heat losses, dust accumulation,
and protection needed at the plant. The data used for this work are
coming from both the plant and the GAST meteo station.

INSOLATION CONDITIONS FOR 1982, 1983, AND 1984

Looking at clear days, the daily beam irradiation values for 1982 show a
low marked difference between summer and winter: the longer summer days
are compensated for by less sky transparency. This effect is not as evident in 1983 and 1984.

The total irradiation (beam irradiance) for the three years was 1730,
2005, and 2020 kWh/m^2 respectively, while the number of irradiance hours
above 300 W/m,^{2} were 2500, 2800, and 2700 hours.

GLOBAL IRRADIATION

A study covering the first six months of 1984 is presented. Approximately 900 kWh/m^2 were received on a horizontal surface, which is the expected value for this region.

INSOLATION DIFFERENCES BEFORE AND AFTER SOLAR NOON

The period of October 1983 - June 1984 was studied and indicates that
during winter months, less clouds are present after solar noon. The
opposite. can be seen in the summer months.

WIND CONDITIONS ON SITE

A study using GAST meteo data for the first six months of 1984 gives the
following conclusions:

- The dominant wind conditions on site are those from the east or west.
- There are east winds every month which is not the case for west winds.
  However, when there are west winds, they may be very strong.

124

ENVIRONMENTAL CONDITIONS - REFLECTIVITY

INTRODUCTION

The SSPS solar facilities use concentration as a method of collecting
solar thermal energy to make electricity. Consequently the
environment and how the environment affects atmospheric transparency,
reflectivity and the rate of reflectivity change is of major interest
in the evaluation of these systems. Considerable effort was expended
at the SSPS site to measure the elements of the environment and to
understand those elements which modified the expected performance of
these systems. A lesson learned from this effort was that extensive
measurement of reflectivity provided the result of the environmental
affect and that it was much more difficult to completely understand
the elements that caused the result.

P. Wattiez led the data collection effort, assuring measurement of
humidity, as a function of time of day, mirror position,
mode of operation, wind condition, etc. in the
interest of developing the understanding of what was taking place.
In cooperation with M. Sanchez, Mr Wattiez prepared the report
ENVIRONMENTAL CONDITIONS; IMPACTS ON SOLAR MIRROR REFLECTIVITY
DEGRADATION.

The conclusions are;
    1 - Loss of reflectivity, caused by the environment is dependent
        on: (a) mirror surface (b) mirror history (c) weather
        conditions (humidity) and (d) mirror position.
    2 - Mirror cleaning is a serious requirement for a solar plant.

Comprehensive measurement of the reflectivity of all the mirrors on a
daily basis is a formidable task. M. Sanchez, realizing the enormity
of this task, studied the data that was being collected and based on
the data, developed a method of estimating the reflectivity of the
fields, using a statistical variable , which behaves as a normal
distribution and depends on several defined factors. This work is
reported in A METHOD FOR ESTIMATING THE REFLECTIVITY DISTRIBUTION.

There are no direct conclusions in this report.

ENVIRONMENTAL CONDITION IMPACTS ON SOLAR MIRROR REFLECTIVITY DEGRADATION

Pierre Wattiez and Manuel Sánchez, ITET

SUMMARY

The environmental conditions of the SSPS site have a serious effect on the mirror reflectivity of the solar collectors. There are close to 12000 m^2 of mirror surface at this site, the reflectivity of which is affected by soiling, and to a certain extent, corrosion and delamination is caused by this environment.

A reflectivity measurement procedure was defined at the beginning of the operational phase of the project and has stimulated gathering of data on the three main mirror fields, i.e., the heliostat field, the single-axis and dual-axis tracking collector fields. In March 1984, the procedure was extended to the new and third field of the DCS plant.

From weekly measurement, reflectivity variations versus time caused an adjustment of the procedure and later on, development of a mathematical model for estimating average field reflectivity.

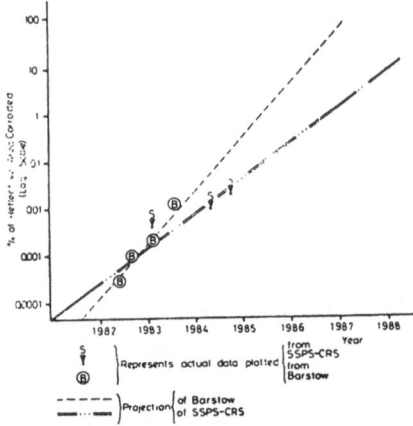

PROJECTED AMOUNT OF CORROSION AT PRESENT
GROWTH RATE (1982)

Continued development of the model made comparisons of the measured and estimated average field reflectivity very satisfactory. A review of the soiling conditions and penetration in the heliostat field, the corrosion problem with the silver layer of the heliostat mirror panel, and a statistical evaluation of its growth through several visual inspections is presented. Based on an estimate growth projection, the viability of the CRS reflective surface could be seriously affected at the end of this decade (1990).

The soiling conditions observed in the DCS collector fields presented some similarities with the heliostat field. Physical and geographical differences cause some differences with the parabolic trough collector average field reflectivity variations. Mirror delaminations were observed and investigated in those fields.

126

The washing experiences gained are reported as a function of their energetic and economical benefits for each mirror field. Suggestions of cleaning improvements are presented based on research results of glass washing techniques.

In conclusion, the soiling impact on the mirror reflectivity observed at the SSPS field is dependent on the physical characteristics of the mirror surface, on the environment conditions, weather conditions, and also on the exposed glass soiling history.

The mirror cleaning is a must for a solar power plant. The efficiency of washing could be improved by the development of an appropriate washing tool. The survey of the mirror field reflectivity will permit an optimization of the solar energy collection on the receiver.

METHOD FOR MEASURING THE REFLECTIVITY DISTRIBUTION

Manuel Sánchez, ITET

SUMMARY

The mean reflectivity of the heliostat field is an important variable for CRS performance and its evaluation. To estimate this parameter on the basis of actual reflectivity measurements on heliostat mirrors is a complex task, however. The problem is how to select a proper set of measurement points and how to extrapolate from the set of measured reflectivities at these points to the rest of the mirror area.

The procedure chosen was to measure 2 points on each of the 93 heliostats. The 2 points are located in the center of the 2 lowest facets in order that the measurements can be made without disturbing plant operation.

In addition, more detailed measurements on some selected heliostats were made in order to study the reflectivity variation over a single facet and the different facets of a given heliostat.

To derive a reflectivity distribution for the whole field, the measured points were used as the fix points for a linear interpolation. This distribution was then assumed to be produced by a combination of the position effect (the effect looked for in the analysis) and a random variation which includes any other causes. The statistical variable defined in this way was assumed to have a normal distribution.

In order to check the validity of the assumptions made, a variance analysis was carried out.

Two minicomputer programs were written to calculate and plot the reflectivity distributions. One for the whole heliostat field, the other for a single heliostat facet. In the case of the single facet, the patterns of reflectivity distribution correlate somewhat with the mechanical construction of the facet. In the case of the heliostat field, the reflectivity distribution cannot be correlated systematically to any parameter, we could imagine.

SOILING

INTRODUCTION

The most important cause of loss of reflectivity is contamination of
the mirror surface, generally a result of contamination from the
atmosphere, refered to here as Soiling. In the interest of
understanding soiling, to determine what material causes loss of
reflectivity and the source of that material, a collection and
analysis program was developed in cooperation with the University of
Seville. V. Ruiz and J. Usero summarized this effort in the report
DUST ANALYSIS.

The conclusions are;
    1 - Chlorides, sulphates and calcium are the major soiling
        agents.
    2 - The amount of soiling agents is not compared to industrial
        zones. (The SSPS site is in a very rural zone)

Understanding what the agents that cause the loss of reflectivity are
leads to attempts at reducing the affect of these agents. Following a
detailed investigative effort on this subject at the site, I. Susemihl
expanded the study area and in conjunction with some industries,
surveyed the present possibilities and summarized them in the report
SOILING EFFECTS; COATING INVESTIGATION ON GLASS AND MIRROR SURFACES.

The conclusions are:
    1 - An effective antisoiling mirror surface should be smooth,
        hard and hydrophobic.
    2 - Intensive rain is an effective cleaning agent.
    3 - Wind is not an effective cleaning agent.
    4 - Light rain may be a cleaning agent on hydrophobic surfaces.

DUST ANALYSIS

V. Ruiz and J. Usero, University of Seville

SUMMARY

Evaluation work performed on thermal solar plants, emphasizes the importance of reflectivity losses on electric plant efficiency. The most evident cause for this is the dust deposited on the mirrors. This paper reports on an analysis performed at the University of Seville on samples taken from the solar platform in Almería. The amount of sedimentable dust, the soluble and insoluble fractions were measured, and chemical composition of this dust was analyzed. In addition, some measurements were made on dust in suspension at the site.

Neither the measurements of the quantity of dust, nor those of the reflectivity loss were made with sufficient periodicity to establish numerical correlations. However, from the available data it has been found that the time evolution of the amount of sedimentable material during the period presents a relation to the loss of reflectivity at the SSPS Project. The study conclusions can be summarized as follows:

- The levels of soluble, insoluble, and total dust deposits are similar in each of the samples.

- The differences in chemical composition between the samples are very small.

- The chemical composition of the soluble fraction does not vary greatly, at least in the elements which were analysed.

- Chloride, sulphate, and calcium are the major constituents of the soluble fraction; they amount to over 40% of the total soluble fraction.

- The levels of sedimentable material and the suspended particles are not high, compared to industrial zones.

From the first two points, it can be concluded that the levels of sedimentable material in the fields are represented by the average value of the four collector stations.

SOILING EFFECTS: COATING INVESTIGATION ON GLASS AND MIRROR SAMPLES

Ingo Susemihl, Fachhochschule Wedel, Germany

SUMMARY

For concentrating collector systems a high specular reflectivity of the
mirrors is desirable. Scattered light as well as absorbed light means
loss of energy and therefore the cleanliness of the heliostat mirrors is
important. Specular reflectance losses as great as 30% have been observed for mirrors exposed for only a few weeks, hence the investigation
described in the paper shows that it is possible to develop treatments for
the glass surface of the heliostats to minimize soiling and to facilitate
cleaning.

Various coatings (on glass and mirror surfaces) were examined to deter-
mine which materials would produce the greatest reduction of soiling.
This paper presents the results of laboratory investigation (done in the
Fachhochschule Wedel) as well as results of a measurement campaign per-
formed in Almería with a special test stand having 96 treated samples.
Those results can be summarized as follows:

- Surface treatment does not affect initial rate of soil accumulation.
- High relative humidity, dew, or soft rain may function as a natural
  cleaning agent on hydrophobic surfaces.
- Intensive rain (or water spray apparatus) is an effective cleaning
  agent and does not depend on surface material.
- Wind (or air spray) does not function as an effective cleaning agent.
- Conductive surfaces showed no difference in soiling behavior.
- The teflonized surfaces are not hydrophobic after a short period of
  outdoor exposure.
- Rough and/or porous surfaces showed a greater affinity to soil.
- The soiling process is faster during the day than at night.

A desired surface should be: smooth, hard, and hydrophobic. Polysiloxane
coatings meet these requirements at least for several months.

Current investigations with the Fachhochschule Wedel on this topic are
described.

SSPS - CRS BIBLIOGRAPHY

CRS-SSPS SEMIANNUAL REPORTS

Report N⁰	Title	Author	Date
SR II	CRS - Construction Report	M. Becker et al	Mar. 1983
SR IV	CRS - First Period of Operation	W. Bucher	May 1984
SR V	CRS-ASR Construction Experience Report	J. Hansen	Sep. 1984

CRS-SSPS TECHNICAL REPORTS

Report N⁰	Title	Date
1/79	Heliostat Field and Data Acquisition Subsystem for CRS (by Martin-Marietta)	Dec. 1979
2/79	CRS-Heliostat Field, Interface Control and Data Acquisition System (by McDonnell Douglas)	Dec. 1979
2/80	Analysis of Special Hydraulic Effects in the SHTS Piping System (by Belgonucleaire)	Nov. 1980
3/80	Redesign of the CRS - Almería Receiver Aperture and Comparison of INTERATOM and MMC Heliostat Field Performance Calculations (by INTERATOM)	Nov. 1980
3/81	CRS Instrumentation Review (by Belgonucleaire)	Jun. 1981
5/81	Device for the Measurement of Heat Flux Distributions (HFD) Near the Receiver Aperture Plane of the Almería CRS Solar Power Station (by DFVLR)	Nov. 1981
1/82	SSPS Workshop on Functional and Performance Characteristics of Solar Thermal Pilot Plants, Part II, Results of the Tower Facilities Session (by M. Becker, DFVLR)	Jul. 1982
2/82	Concentrated Solar Flux Measurements at the IEA SSPS Solar Central Receiver Power Plant, Tabernas, Almería (Spain) (by G. von Tobel, C. Schelders, M. Real, EIR)	Apr. 1982
3/82	Effect of Sunshape on Flux Distribution and Intercept Factor of the Solar Tower Power Plant at Almería (by G. Lemperle, DFVLR)	Sep. 1982
3/83	The Advanced Sodium Receiver (ASR) - Topic Reports (by AGIP Nucleare and Franco Tosi)	May 1983
4/83	CRS-Midterm Workshop (edited by M. Becker, DFVLR)	Jun. 1983
5/83	Investigations and Findings Concerning the Sodium Tank Leakages (edited by W. Bucher, DFVLR)	Jul. 1983
7/83	Thermal Losses of the Sodium Storage Vessels of the Central Receiver System (by H. Jacobs, SSPS-ITET)	Nov. 1983
1/84	Executive Summary - IEA/SSPS - CRS Workshop (April 1983) (by C.S. Selvage, SSPS-ITET)	Mar. 1984

Report NO	Title	Date
1/81	Tabernas Meteo Data Analysis Based on Evaluated Data Prepared by the SSPS O.A. (by Belgonucleaire)	Jun. 1981
4/81	International Energy Agency Small Solar Power Systems (SSPS) Project Review (January 1981) (by A.F. Baker, SANDIA)	Jul. 1981
6/81	Determination of the Spectral Reflectivity and the Bidirectional Reflectance Characteristics of Some White Surfaces (by DFVLR)	Dec. 1981
2/83	FH-PTL Wedel Reflectometer, Type 02-1 No. 3 Final Report and Report on the Test Program (by G.Lensch, K. Brudi, P. Lippert, Fachochschule Wedel)	Mar. 1983

CRS-SSPS INTERNAL REPORTS

Subject: Heliostat Field

Report NO	Title	Author	Date
R-1/81	Test Report: Heliostat Reflectance Spot Check	R.P.Stromberg	24.7.81
R-9/81	Technical Changes Incorporated in the MMC Heliostat Controllers (HAC) after Acceptance	W. Grasse	26.8.81
R-17/81	Heliostat Reflectance Estimate	R.P.Stromberg	24.9.81
R-27/81	Heliostat Status Report for Month of October	E. Madigan	17.11.81
R-43/82	Belgian Heliostat Evaluation since September 1981 until March 1982	T. van Steen-berghe/ P. Wattiez	11.5.82
R-76/82	Reflectivity Measurements over the Heliostat Field: 15.5.82-3.8.82	P. Wattiez	13.9.82
R-86/82	F.M.C. (2.9-14.10.82) Heliostat Field Reflectivity Behavior	P. Wattiez	8.11.82
R-94/82	Heliostat Field Reflectivity	LR/MS	13.12.82
R-26/83	Heliostat Field Investigation Program	P. Wattiez	16.5.83
R-27/83	Investigation of Water Contained in Heliostat Modules	P. Wattiez	13.6.83
R-29/83	Wind Studies in Respect of Mirror Field Stowage	M. Loosme	8.8.83
R-6/84	Mirror Corrosion and Heliostat Condition	M. Sánchez/ J. Ramos	30.3.84
R-12/84	Simulation of Tracking Errors	M. Blanco	5.4.84
R-14/84	Status of Heliostat Field Alignment	J. Ramos	24.4.84
R-24/84	Heliostat Flux Distribution, Weekly Report (week nO 25)	A. Cuadrado	29.6.84
R-25/84	Heliostat Flux Distribution, Weekly Report (week nO 26)	A. Cuadrado	3.7.84
R-27/84	Heliostat Flux Distribution, Weekly Report (week nO 28)	A. Cuadrado	24.7.84
R-28/84	Heliostat Flux Distribution, Weekly Report (week nO 29)	A. Cuadrado	30.7.84

Subject: Heliostat Field (continued)

Report NO	Title	Author	Date
R-29/84	A Mathematical Model for Estimating Average Heliostat Field Reflectivity	M. Sánchez/ P. Wattiez	10.5.84
R-30/84	Heliostat Field Reflectivity	M. Sánchez/ L. Ruiz	17.7.84
R-31/84	Heliostat Flux Distribution, Weekly Report (week no 30)	A. Cuadrado	6.8.844
R-32/84	Reflectivity Measurement and Meteo Data Collection from Feb.20 to Apr. 10, 1984	H. Winter	3.5.84
R-35/84	Heliostat Flux Distribution, Weekly Report (week no 32)	A. Cuadrado	21.8.84
R-41/84	Multiple Seasonal Model for Predicting the Average Reflectivity (Computer Program Application to SSPS-CRS Heliostat Field)	M. Sánchez	10.5.84
R-48/84	On Site Calculations with the HELIOS Code	H. Jacobs	25.10.84

Subject: Receiver

Report NO	Title	Author	Date
R-22/81	Inputs for the Design of the Advanced Sodium Receiver	INTERATOM	16.10.81
R-56/82	CRS Receiver Reradiation and its Measurement Calculation Note	T. van Steenberghe	28.6.82
R-58/82	CRS Receiver - Reradiated Flux Distribution (RFD) Measurement System	T. van Steenberghe	7.782
R-70/82	Heat Flux Distribution Measurement System: Special Versions of the Software for the Kendall Radiometer	T. van Steenberghe	2.9.82
R-77/82	Receiver Efficiency Calculation Method CRS Fall Equinox Measurement Campaign 1982	J. Kraabel	16.9.82
R-78/82	Almería Steam Generator. Part Load Characteristic with Reduced Steam Pressure	Sulzer AG	24.9.82
R-85/82	Solar Absorptance Measurements on the First CRS Receiver Tubes	T. van Steenberghe	29.10.82
R-1/83	Measurement and Evaluation of Temperature Inside the CRS Receiver	F. Gaus	25.2.83
R-12/83	Test Program for Measurement and Evaluation of the Receiver Tube Bundle Movements with Displacement Transducers	F. Gaus	9.3.83
R-23/83	Solar Hemispherical Absorptance of the CRS First Receiver Ceramic Wall	T. van Steen-	4.8.83
R-24/83	ITET Participation in the ASR-Functional Test	H. Jacobs/ M. Pescatore	10.8.83
R-28/83	Convection Losses of the Sulzer Receiver	H. Jacobs	24.8.83

Subject: Receiver (continued)

Report No	Title	Author	Date
R–33/83	Measurement of the Distance Between Receiver Tubes and a Reference Line, and Control of the Central Support in the Sulzer Receiver	F. Gaus	7.9.83
R–37/83	ASR Incident on September 7, 1983	A.DeBenedetti A. di Meglio J. Hansen	19.9.83
R–11/84	Steady-State and Transient Conditions in a Tube with Incident Thermal Flux and Inlet Condition Assigned: Program ASRTTHB (ASR Tube Thermal Balance)	A.DeBenedetti	12.4.84
R–17/84	A Comparison of the SULZER Cavity Receiver and the FRANCO-TOSI External Receiver	M. Pescatore	22.5.84
R–19/84	Receiver Vent Valves Control System Modification	A. Cuadrado	30.5.84
R–21/84	Preliminary Results on the Performance of the SULZER Cavity Receiver and the FRANCO-TOSI External Receiver	C.S. Selvage/ H. Jacobs	26.6.84
R–33/84	Heliostat Field – Receiver Overall Simulation	A.DeBenedetti	6.8.84
R–34/84	ASR Temperature Increase at Central Panel Outlet	A.DeBenedetti J.G. Martín	6.8.84
R–37/84	ASR Ceramic Border Temperatures Versus Incident Radiation	A. Cuadrado J. Ramos	28.8.84
R–38/84	THERESA: A Thermal Analysis Code for Billboard Receivers	A.DeBenedetti	17.8.84
R–46/84	Sulzer Receiver Transient Response Data During Period March–April 1983	N. Gregory	11.10.84
R–47/84	Available Point and Class Summary Data: Sulzer Receiver	N. Gregory	18.10.84

Subject: Tank

Report No	Title	Author	Date
R–5/82	Determination of Sodium Tank's Volume as a Function of Level	C. Gómez	4.2.82
R–26/82	Failure and Repair of the Cold Storage Vessel LK01 BB01 – Preliminary Incident Report	D. Stahl/ INTERATOM	26.3.83
R– 68/82	Sodium Leakage at the SSPS–CRS Cold Storage Tank	W. Bucher	11.8.82
R–84/82	Thermal Shock as a Possible Cause of Sodium Leaks	J.G. Martín	21.10.82
R–4/83	Thermal Losses of the Sodium Storage Vessels	H. Jacobs	9.2.83

Subject: Power Conversion System

Report NO	Title	Author	Date
R-32/82	Test Program for Improving the Load Change Behavior in the Steam Generator Casing	H. Jacobs	5.4.82
R-44/82	CRS PCS Gross Efficiency from November 11, 1981 to December 12, 1981	C. Gómez/ R. Carmona	14.5.82
R-63/82	CRS Steam Motor Alternatives-Discussions held at SSPS, Tabernas 15.-16.7.1982	J. Hansen	22.7.82
R-66/82	Steam Generator Load Change Behavior Improvement	H. Jacobs	29.7.82
R-73/82	Spilling Steam Motor 10 Day Test	J. Hansen/ F. Martinez	23.8.82
R-79/82	Supplement to the Spilling Steam Motor Test Report NO R-73/82	J. Hansen	27.9.82
R-10/83	Review of Spilling Performance Status Report at Final Review on CRS Repair, 24.3.83	J. Hansen	22.3.83

Subject: Miscellaneous

Report NO	Title	Author	Date
R-3/81	Information Concerning the Status of CRS Performance Determination	M. Becker	4.8.81
R-11/81	Receiver Platform Incident due to Power Failure	W. Grasse	1.9.81
R-11a/81	Final Incident Report on Receiver Platform Incident	E. Madigan/ F. Martinez/ D. Stahl	10.12.81
R-23/81	Reporting System	W. Grasse/ P. Wattiez/ A. Baker	29.10.81
R-30/81	Input for SSPS Status Report given at the CRS-Department of Energy Annual Meeting (October 13-16, 1981)	E. Madigan	17.11.81
R-21/82	Revision of CRS-DAS Calculation	C. Gómez	2.3.82
R-48/82	Recommendations for CRS/HAC after visit of Ed Madigan	E. Madigan	1.6.82
R-49/82	Program of the CRS Measurement Campaign around Fall Equinox - Status June 1982	M. Becker	22.6.82
R-83/82	CRS Daily Summary, Description, and Comments	P. Wattiez	27.10.82
R-2/83	Comparison of the SSPS-CRS Heat Flux Measurements and Corresponding Theoretical Predictions	Dr. Kiera/ INTERATOM	25.1.83
R-5/83	SSPS Stage 3 as Promoting Program of the Solar Energy Diffusion	P. Wattiez	3.2.83
R-6/83	CRS Fall Equinox Measurement Campaign	T. van Steen-berghe	16.1.83
R-13/83	HFD Repair Report	J. Ramos	24.3.83

Subject: Miscellaneous (continued)

Report N°	Title	Author	Date
R-17/83	Status Report of CRS / SHTS	F. Ruiz	19.5.83
R-18/83	SOLTES Modelling of the SSPS Plants at Sandia National Laboratories Mission Report	T. von Steenberghe	17.5.83
R-31/83	Status of CRS Before Start-Up ASR Operation	P. Laguía	22.8.83
R-36/83	CRS Plant Operation and its Documentation with Data Tape from November 16, 1981 to August 31, 1983	M. Pescatore	5.9.83
R-3/84	Modifications to the Cold Sodium Pump Control	J. Ramos	20.1.84
R-16/84	Water Physical Property Package	A.DeBenedetti	3.5.84
R-36/84	Trace Heating Consumption	A. Cuadrado	27.8.84
R-42/84	IEA/SSPS Calibration Report- Calibration of Relevant Measuring Sensors	A. Brinner	27.9.84
R-44/84	An Operational Usage Factor for the CRS Plant	N. Gregory	2.10.84
R-45/84	Available Data and Approximation of the Average CRS Plant Start-up Time	N. Gregory/ H. Jacobs	2.10.84
R-51/84	Progress on Lightning Protection Improvements for CRS Heliostat Field (2	J. Ramos	8.10.84

Subject: CRS-DCS Common

Report N°	Title	Author	Date
R-2/81	Measurement Device Test for Solar Insolation	G. von Tobel	28.7.81
R-5/81	Plant Optimization Phase (POP) Test Requirements	B. Wilson/ R.P.Stromberg	10.8.81
R-5a/81	Idem.	A. Baker/ W. Wilson/ R.P.Stromberg	28.9.81
R-5b/81	Idem.		Oct.81
R-5c/81	POP Test Requirements DCS System Tests (new Tests N° 34-37) DCS Updated POP Test Listings CRS Updated POP Test Listings	P. Wattiez	5.11.81
R-6/81	Cost of Electricity	A. Rieger	17.8.81
R-10/81	Cooling Towers for Future Solar Power Plant Projects	A. Rieger	27.8.81
R-12/81	Report of Damages caused by the 2nd Flooding of SSPS Plant, Tabernas, Almería	P. Heintzelmann	1.9.81
R-14/81	Mirror Reflectance Study, SSPS Projects, Tabernas, Spain	R.P.Stromberg	15.9.81
R-16/81	Recommendations for Improvement of the 25 kW Grid Behavior	A. Rieger	23.9.81
R-18/81	Reflectivity Measurement Corrections	R.P.Stromberg	30.9.81
R-19/81	Optimization of the Time Period Between Mirror Washing	A. Rieger	2.10.81
R-21/81	Comparison of Solar Radiation Measurement Devices	G. von Tobel	15.10.81
R-3/82	Measurement Program of Diffuse Component on Clear Skies	A. Sevilla	29.1.82

Subject: CRS-DCS Common (continued)

Report N⁰	Title	Author	Date
R-23/82	Solar Heating for the SSPS Office Buildings	A. Sevilla	15.3.82
R-27/82	Comparitive Tests on Two Portable Reflectometers D&S-15R + FH-PTL-02.1	P. Wattiez	30.3.82
R-28/82	Reflectivity Measurement Procedure	P. Wattiez	31.3.82
R-41/82	A General Purpose Polynomial Regression Program	T. van Steenberghe	11.5.82
R-45/82	Plant Safety Meeting	M. Loosme	29.4.82
R-46/82	Results of Reflectivity Variation Over the SSPS Mirror Fields	P. Wattiez	27.5.82
R-52/84	Insolation at the SSPS Site in Tabernas (1981 and 1982 till May)	P. Wattiez	17.6.82
R-60/82	Lightning Protection of the SSPS Plant	M. Loosme	13.7.82
R-65/82	Reflectance Properties of Mirror Modules from the IEA Project	R.B. Petit/ A.R. Mahoney/ SNLA	26.7.82
R-71/82	Cleaning Cost of the SSPS Fields	P. Wattiez	10.9.82
R-72/82	SSPS Field Cleaning Frecuency	P. Wattiez	11.8.82
R-75/82	Intermediate Report on the Photovoltaic Experiment at the IEA/SSPS Site in Tabernas (Spain) August 1982	P. Toggweiler	8.9.82
R-11/83	Computer Plotting as Evaluation Tool	M. Andersson	9.3.83
R-20/83	US Trip of Dr. Kalt March 1983 Travel Report	A. Kalt	Apr. 1983
R-22/83	1983 Solar Ephemeridis for the Plataforma Solar, Tabernas (Almería)	T. van Steenberghe	8.7.83
R-25/83	CRS and DCS Common Parasitic Sources	P. Wattiez	23.2.83
R-30/83	Passive Solar Retrofit of the SSPS Building Office	A. Sevilla	26.8.83
R-34/83	Energy Calculations of an Office Building Using the DOE-2.1a	L. Chien-Kuo You/ J.G. Martín	23.9.83
R-41/83	Monthly Direct Insolation Data	L. Castillo	7.10.83
R-5/84	Earthing System Inspection	B. Calatrava	26.3.84
R-8/84	Report of Work Performed at the Small Solar Power Systems (SSPS) Site 1.7-31.8.83	W. Banhardt/ S. Lauties	23.3.84
R-10/84	Safety Meeting 2/84	B. Calatrava	8.4.84
R-18/84	Safety Meeting	B. Calatrava	22.5.84
R-20/84	Safety Meeting	B. Calatrava	6.6.84
R-40/84	Wind Conditions on the SSPS Site	L. Castillo	27.9.84
R-50/84	The IEA/SSPS Office Building Cooling and Heating Loads: Effect of Retrofit Options	L. Chien Kuo You/ J. Martín	12.7.84

SSPS - DCS BIBLIOGRAPHY

DCS-SSPS SEMIANNUAL REPORTS

Report No	Title	Author	Date
SR I	DCS - Construction Report	A. Kalt et al	Nov. 1982
SR III	DCS - First Year of Operation	A. Kalt	Jul. 1983
SR IV	DCS - Supplement Construction Report	J. Hansen	Nov. 1984
Sr VI	DCS -Supplement Construction Experience Report	J. Hansen	Nov. 1984

DCS-SSPS TECHNICAL REPORTS

Report No	Title	Date
1/80	Collector Qualification Tests for the IEA 500 kW$_e$ Distributed Collector System (by SANDIA and DFVLR)	Jul. 1980
2/81	DCS Instrumentation Review (by Belgonucleaire)	Jun. 1981
1/82	SSPS Workshop on Functional and Performance Characteristics of Solar Thermal Pilot Plants, Part I, Results of the DCS-Plant Sessions (by A.Kalt,DFVLR)	Apr. 1982
1/83	DCS-Midterm Workshop Proceedings (edited by A.Kalt, J. Martín)	Feb. 1983
6/83	First Year Average Performance of the SSPS DCS Plant (by T. van Steenberghe, SSPS-ITET)	Jul. 1983
2/84	SSPS-Distributed Collector System, Proceedings of the International Workshop "The First Term" (Dec. 6-8, 1983, tabernas, Spain)	May 1984
3/84	SESAM-DCS, A Computer Code for Solar System Modelling, Part I - Analysis Report, Part II - How to Use, (by J. Fabry, H. Richel, H. Lamotte, M. Vereb, P. Brusselaers)	Mar. 1984
4/84	The Control of Large Collector Arrays: The SSPS Experience (by R. Carmona and J.G. Martín)	Jun. 1984
5/85	SSPS-DCS Plant Performance, "The Stair-Step" (by P. Wattiez, J.G. Martín, M. Andersson)	Jul. 1984

DCS/CRS TECHNICAL REPORTS (common)

Report No	Title	Date
1/81	Tabernas Meteo Data Analysis Based on Evaluated Data Prepared by the SSPS O.A. (by Belgonucleaire)	Jun. 1981
4/81	International Energy Agency Small Solar Power Systems (SSPS) Project Review (January 1981) (by A.F. Baker, SANDIA)	Jul. 1981
6/81	Determination of the Spectral Reflectivity and the Bidirectional Reflectance Characteristics of Some White Surfaces (by DFVLR)	Dec. 1981
2/83	FH-PTL Wedel Reflectometer, Type 02-1 No. 3 Final Report and Report on the Test Program (by G.Lensch, K. Brudi, P. Lippert, Fachhochschule Wedel)	Mar. 1983

DCS-SSPS INTERNAL REPORTS

Subject: DCS Plant

Report No	Title	Author	Date
R-25/81	DCS Punch List Inspection Report	M. Loosme	9.11.81
R-26/81	Report on DCS POP Activities from Sept. 21 until Nov. 21, 1981	T. van Steen-berghe	17.11.81
R-28/81	Operation of DCS for Maximum Electrical Energy Generation	H. Dehne	17.11.81
R-29/81	150 Kw 1981 Annual Report Excerpts	H. Dehne	17.11.81
R-34/81	Comments on DCS Performance Evaluation by P. Pezuela (10.9.81) and Dr. Kemper (18.9.81) Calculations of DCS Performance Evaluation	C. Gómez	18.11.81
R-36/81	Evaluation of DCS Acceptance Test Data (Measurement of Sept. 7, 1981)	A. Kalt	5.10.81
R-7/82	Evaluation Report on DCS + CRS Activities during Optimization Phase (Sept. - Dec. 1981)	P. Wattiez	10.2.82
R-37/82	Evaluation Report on DCS + CRS Operational Activities from Sept. 21, 1981 until Apr. 15, 1982	P. Wattiez/ T. van Steen-berghe	29.4.82
R-42/82	Behavior Improvement of the DCS Plant	P. Wattiez	11.5.82
R-51/82	DCS Plant Performance Measurements around Summer Solstice	A. Kalt	16.6.82
R-55/82	Daily Summary - Description of the Daily Filled Form for the DCS	P. Toggweiler	24.6.82
R-59/82	Detailed Measurements at the DCS Plant around Summer Solstice 1982	A. Kalt	8.7.82
R-67/82	Determination of the Feasibility of transferring CRS energy DCS-PCS	P. Wattiez	5.8.82
R-80/82	Concept for the Routine Operation Phase of the SSPS DCS Plant- 2nd Draft	A. Kalt	14.9.82
R-87/82	Performance Corrections	J.G. Martín	9.11.82
R-88/82	Balance of One Year of Operation with DCS-SSPS	P. Wattiez	13.11.82
R-90/82	First Year Average Performance of the SSPS-DCS Plant	T. van Steen-berghe	9.12.82
R-35/83	Investigation of DCS Parasitic Consumption	P. Wattiez	23.9.83
R-15/84	SSPS-DCS Plant Performance "THE STAIR STEP"	P. Wattiez/ J.G. Martín/ M. Andersson	30.4.84
R-22/84	DCS Losses	H. Jacobs/ M. Andersson/ R. Carmona/ M. Pescatore	22.5.84
R-39/84	HELIOS Modifications for Trough Mirrors	A. DeBenedetti M. Blanco	17.8.84
R-49/84	SESAM-DCS	M. Andersson	29.10.84

DCS-SSPS INTERNAL REPORTS (continued)

Subject: Collector Fields

Report No	Title	Author	Date
R-13/81	Estimate of DCS Collector Output Under Acceptance Conditions (Sep.7,81)	R.P.Stromberg	15.9.81
R-15-81	Estimate of Mirror Reflectance, Acceptance Test Dates, DCS System	R.P.Stromberg	16.9.81
R-20/81	Collector Cleaning and Soiling Studies	H. Dehne	15.10.81
R-24/81	A Calculation of Field Losses for SSPS-DCS Fields	R.P.Stromberg	11.11.81
R-11/82	Comparison Between Solar Energy Offered to ACUREX Field and to MAN Field	C. Gómez	16.2.82
R-22/82	Stationary Piping Thermal Losses in ACUREX and MAN Fields with Collectors in Desteer	C. Gómez	4.3.82
R-35/82	Shadow Effects in DCS Collector Field	C. Gómez	19.4.82
R-74/82	Energy Losses in a One-Line Focus Parabolic Concentrating Collector without Incidence angle Effects	C. Gómez	3.9.82
R-91/82	Solar Hemispherical Absorptance Measurements on SSPS-DCS Receiver Tube Selective Coatings	T. van Steenberghe/ A. Sevilla	9.12.82
R-92/82	Collector Field Soiling Conditions	P. Wattiez	11.1.82
R-93/82	DCS Collector Field Behavior over the Period 21.10.81 to 21.10.82	P. Wattiez	26.11.82
R-3/83	Correction of Collected Field Measurement Efficiency to Design Conditions	J.G. Martín/ A. Kalt	9.2.83
R-7/83	SSPS Project Mirror Modules Status Report	P. Laguía/ F. Martinez	11.3.83
R-15/83	DCS Automatic Flow Control Valves	P. Laguía	12.4.83
R-16/83	Temperature Drop in the DCS Storage Tank Effects of Improved Insulation	M. Andersson	2.5.83
R-19/83	Insolation in ACUREX Field - Tilt Angle Optimization	R. Carmona/ L. Castillo	23.2.83
R-21/83	Baeltz Automatic Valve Positioner	P. Laguía	6.6.83
R-39/83	Proposal of Oil Field Bypass Procedure of Operation	R. Carmona	13.5.83
R-40/83	Pumping Power Consumption of the Main DCS Pumps	T. van Steenberghe R. Carmona	29.7.83

Subject: ACUREX Fields

Report NO	Title	Author	Date
R-4/82	Direct Normal Insolation in ACUREX Field	C. Gómez	4.2.82
R-12/82	A Method of Evaluating the Possible Effects fo Superheating in Selective Surfaces in ACUREX Collectors	C. Gómez	16.2.82
R-13/82	Status of ACUREX Absorbers Coating	A. Sevilla	18.2.82
R-16/82	Examination of Thin Glass Panel De-lamination (14 + 15.12.81)	J.W. Jacob/ GLAVERBEL	24.2.82
R-20/82	Efficiency of ACUREX Model 3001-03 Parabolic Trough Concentrating Solar Collector as function of solar irra-diance and average receiver fluid temperature above ambient	C. Gómez	1.3.82
R-36/82	ACUREX Absorber Coating Testing	A. Sevilla	28.4.82
R-47/82	Sun Position and Incidence Angle Cal-culations for the ACUREX Field	T. van Steen-berghe	2.6.82
R-50/82	Determination of Absorptance and Emi-ttance Degraded ACUREX Receiver Tube Samples	G. Görler/ A. Kalt	16.6.82
R-54/82	Determination of Absorptance and Emi-ttance Degraded ACUREX Receiver Tube Samples	ACUREX Corp.	23.6.82
R-69/82	Solar Absorptance Measurements on ACUREX Receiver Tube	T. van Steen-berghe/ A. Sevilla	2.9.82
R-81/82	ACUREX Field Mirror Delamination Sta-tus	P. Laguía	5.10.82
R-13/84	Status of ACUREX Field Mirrors	A. Cuadrado	24.4.84
R-23/84	Status of ACUREX Field Mirrors (5.5.84)	A. Cuadrado	22.6.84
R-43/84	Estimate of Oil Flow Rate in the ACUREX Field	R. Carmona	1.10.84

Subject: MAN Field

Report NO	Title	Author	Date
R-31/81	Status of MAN Field Temperature Con-trol System	H. Dehne	17.11.81
R-6/82	MAN Testing (Jan./Feb. 1982)	J. Hansen	8.2.82
R-14/82	Configuration and Constants Implemen-ted in MAN Field Control	H. Dehne	24.2.82
R-15/82	Warranty Work in the MAN Collector Field 29.1 - 5.2.1982	P. Kemper	24.2.82
R-33/82	MAN Activities on SSPS Site from 29.3.82-7.4.82	J.P. Kemper/ MAN	12.4.82
R-38/82	MAN Field Temperature Control System	H. Dehne	30.4.82
R-2/84	Final Revision of the Earthing Net-work of the new MAN Field	B. Calatrava	11.2.84
R-9/84	Reflectivity Measurement Procedure for the MAN II Field	P. Wattiez	4.4.84

Subject: Storage Subsystem

Report NO	Title	Author	Date
R-35/81	DCS Thermal Energy Storage Tank Demonstration Report	C. Gómez	19.11.81
R-39/81	Method for Evaluating the Mass and the Energy Stored in the Thermocline Tank	C. Gómez	10.12.81

Subject: Power Conversion Ssytem (PCS)

Report NO	Title	Author	Date
R-32/81	Improvement Work in the DCS PCS which was accomplished during Stal-Laval's presence on Site between Oct.28 and Nov. 13, 1981	J. Hansen	17.11.81
R-38/81	PCS Start-Up Procedure	J. Hansen	4.12.81
R-40/81	Attainable Steam Temperature After Steam Generator Modification	A. Kalt	1.12.81
R-1/82	DCS Steam Generator, Corrosion Protection during Idle Time	J. Hansen	19.1.82
R-9/82	Gross and Net Efficiency of PCS-DCS Plant	C. Gómez	8.2.82
R-10/82	Thermal Losses in PCS-DCS	C. Gómez	16.2.82
R-24/82	DCS Boiler Water Treatment	J. Hansen	16.3.82
R-34/82	DCS Steam Generator Inert Gas System	J. Hansen	14.4.82
R-53/82	Efficiency and Thermal Losses in PCS-DCS Plant in June 1982	R. Carmona	21.6.82
R-61/82	Condenser Level Control	P. Laguía	14.7.82
R-82/82	DCS Steam Generator Leakage, Repair and Investigation September 1982	J. Hansen	7.10.82
R-42/83	Power Conversion System Performance for the SSPS Distributd Collector Systems	R. Carmona/ H. Jacobs	14.10.83

Subject: Master Control System/Data Acquisition System

Report NO	Title	Author	Date
R-33/81	DCS Plant - MCS/DAS Suggested Modifications	T. van Steenberghe	18.11.81
R-8/82	Verification of DCS/DAS Virtual Calculations	T. van Steenberghe/ C. Gómez	16.2.82
R-17/82	MCS/DAS Software Additions	Electrowatt	26.2.82
R-29/82	Proposal to Improve the Use of the File Capacity in the DCS MCS/DAS	P. Toggweiler	1.4.82
R-32/83	Visit to Dornier Systems (5+6, Sept.) Maintenance of MAN-Electronic Boxes	J. Ramos	16.9.83
R-39/82	DCS/DAS Virtual Point Calculation - Actual Implementations and Recommendations for Improvements	P. Toggweiler	5.5.82
R-64/82	Specifications for Changes in the DCS MCS/DAS Software	P. Toggweiler	26.7.82

Subject: DCS-Auxiliaries

Report No	Title	Author	Date
R-19/82	DCS Operation Evaluation: Water Analysis: Recommendation for Further Operations	J. Hansen	3.3.82
R-30/82	DCS - Uninterruptable Power Supply Repair Work March 1982	P. Coppee	5.4.82
R-31/82	DCS Emergency Generator - Repair Work March 1982	J. Asensio/ Tec. Reun.	5.4.82
R-62/82	Calibration of DCS Pyrheliometer	R. Carmona	23.7.82

Subject: DCS-CRS Common

Report No	Title	Author	Date
R-2/81	Measurement Device Test for Solar Insolation	G. von Tobel	28.7.81
R-5/81	Plant Optimization Phase (POP) Test Requirements	B. Wilson/ R.P.Stromberg	10.8.81
R-5b/81	Idem.		Oct.81
R-5c/81	POP Test Requirements DCS System Tests (new Tests No 34-37) DCS Updated POP Test Listings CRS Updated POP Test Listings	P. Wattiez	5.11.81
R-6/81	Cost of Electricity	A. Rieger	17.8.81
R-10/81	Cooling Towers for Future Solar Power Plant Projects	A. Rieger	27.8.81
R-12/81	Report of Damages caused by the 2nd Flooding of SSPS Plant, Tabernas, Almería	P. Heintzelmann	1.9.81
R-14/81	Mirror Reflectance Study, SSPS Projects, Tabernas, Spain	R.P.Stromberg	15.9.81
R-16/81	Recommendations for Improvement of the 25 kW Grid Behavior	A. Rieger	23.9.81
R-18/81	Reflectivity Measurement Corrections	R.P.Stromberg	30.9.81
R-19/81	Optimization of the Time Period Between Mirror Washing	A. Rieger	2.10.81
R-21/81	Comparison of Solar Radiation Measurement Devices	G. von Tobel	15.10.81
R-3/82	Measurement Program of Diffuse Component on Clear Skies	A. Sevilla	29.1.82
R-23/82	Solar Heating for the SSPS Offices	A. Sevilla	15.3.82
R-27/82	Comparitive Tests on Two Portable Reflectometers	P. Wattiez	30.3.82
R-28/82	Reflectivity Measurement Procedure	P. Wattiez	31.3.82
R-41/82	A General Purpose Polynominal Expression Program	T. van Steenberghe	11.5.82
R-45/82	Safety Plan	M. Loosme	29.4.82
R-46/82	Results of Reflectivity Variation Over SSPS Mirror Fields	P. Wattiez	27.5.82
R-52/84	Insolation at the SSPS Site in Tabernas (1981 and 1982 till May)	P. Wattiez	17.6.82

Subject: DCS-CRS Common (continued)

Report NO	Title	Author	Date
R-60/82	Lightning Protection of the SSPS Plant	M. Loosme	13.7.82
R-65/82	Reflectance Properties of Mirror Modules from the IEA Project	R.B. Petit/ A.R. Mahoney/ SNLA	26.7.82
R-71/82	Cleaning Cost of the SSPS Field	P. Wattiez	17.8.82
R-72/82	SSPS Field Cleaning Frequency	P. Wattiez	17.8.82
R-75/82	Photovoltaic Experiment at the IEA/ SSPS Site	P. Toggweiler	8.9.82
R-11/83	Computer Plotting as Evaluation Tool	M. Andersson	9.3.83
R-22/83	1983 Solar Ephemeredis for the Plataforma Solar	T. van Steenberghe	12.7.83
R-25/83	CRS and DCS Common Parasitic Sources	P. Wattiez	23.8.83
R-30/83	Passive Solar Retrofit of the SSPS Building Office	A. Sevilla	26.8.83
R-34/83	Energy Calculations of an Office Building Using the DOE-2.1a	L. Chein-Kuo You/ J.G. Martín	23.9.83
R-41/83	Monthly Direct Insolation	L. Castillo	7.10.83
R-5/84	Earthing System Inspection	B. Calatrava	26.3.84
R-8/84	Report of Work Performed at the Small Solar Power Systems (SSPS) Site 1.7-31.8.83	W. Banhardt/ S. Lauties	23.3.84
R-10/84	Safety Meeting 2/84	B. Calatrava	8.4.84
R-18/84	Safety Meeting	B. Calatrava	22.5.84
R-20/84	Safety Meeting	B. Calatrava	6.6.84
R-40/84	Wind Conditions on Site	L. Castillo	27.9.84

ABBREVIATIONS

ACU   ACUREX Corporation, manufacturer of the one-axis tracking collector

ASR   Advanced sodium receiver, type billboard - external

CRS   central receiver system

DCS   distributed collector system

DFVLR   Deutsche Forschung -und Versuchsanstalt für Luft -und Raumfahrt e.V.

HFS   heliostat field subsystem

HTS   heat transfer system

IEA   International Energy Agency

ITET   International Test and Evaluation Team

kWh   kilowatt hours, a measure of electrical energy. The product of kilowatts of electrical power applied to a load times the hours it is applied. One kWh is equivalent to 3,413 BTU of heat energy.

MAN   Maschinenfabrik Augsburg - Nürnberg Akteingesellschaft

OA   Operating Agent of the project, entrusted to the DFVLR

PCS   power conversion system

POA   Plant Operation Authority, entrusted to Sevillana

SHTS   sodium heat transfer system

SSPS   Small Solar Power Systems

SOLAR TERMINOLOGY

Absorber
: The blackened surface of a collector which absorbs solar radiation and converts it to heat energy: a flat black paint is a good absorber (Argue).

Absorptance
: The ratio of absorbed to incident solar radiation (Kreith).

Absorptivity
: The ratio of absorbed radiation by a surface to the total incident radiated energy on that surface.

Active Solar System
: A system in which a transfer fluid (liquid or gas) is circulated through a solar collector where the collected energy is converted, or transferred to energy in the medium.

Aiming Point
: The focalization point of a heliostat on a receiver.

Air Mass
: The ratio of the actual distance traversed through the Earth's atmosphere by the direct solar beam to the depth of the Earth's atmosphere, normal to the surface (McVeign).

Ambient Temperature
: The surrounding air temperature.

Antireflection Coating
: The application of a thin film of dielectric material to a surface to reduce its reflection and to increase its transmission of light or other electromagnetic radiation (Lapedes).

Aperture Area
: The net area of the collector that intercepts radiation (Kreith).

Array
: An assembly of a number of collector elements, or panels, into the solar collector for a solar energy system.

Attenuation	The reduction of radiation flux over a given path length, due to absorption and scattering (McVeigh).
Auxiliary Energy	In solar energy technology, the energy supplied to the heat or cooling load from other than the solar source, usually from a conventional heating or cooling system. Excluded are operating energy, and energy which may be supplemented in nature but does not have the auxiliary system as its origin.
Auxiliary Energy Subsystem	In solar energy technology the Auxiliary Energy System is the conventional heating and/or cooling equipment used as supplemental or backup to the solar system.
Azimuth	The angle between the south-north line at a given location and the projection of the earth-sun line in the horizontal plane (Kreith).
Beam Radiation	Radiated energy received directly, not from scattering or reflecting sources.
Black Body	A term denoting an ideal body which would absorb all and reflect none of the radiation falling upon it (McVeigh).
Bypass Loop	A piping arrangement which bypasses or circles the flow of a heat absorbing medium around rather than through a piece of mechanical equipment (Burt).
Carnot Cycle	A hypothetical cycle consisting of four reversible processes in succession: an isothermal expansion and heat addition, an isentropic expansion, and isothermal compression and heat rejection process, and an isentropic compression (Lapedes).

152

Carnot Efficiency	The efficiency of a Carnot engine receiving heat at a temperature absolute T_1 and giving it up at a lower temperature T_2 (Lapedes).
Collector	A device which absorbs solar radiation and converts it to heat energy (Argue).
Collector Area	The area of a collector which traps the sun (Argue).
Collector Efficiency	The ratio of th energy collected by a solar collector to the radiant energy incident on the collector (Williams).
Collected Solar Energy	The thermal energy added to the heat transfer fluid by the solar collector.
Collector Subsystem	The assembly of components that absorbs incident solar enrgy and transfers the absorbed thermal energy to a heat transfer fluid.
Concentrating Collector	A collector which uses reflective devices or optical lens arrangements to concentrate the sun's rays onto a small collector/absorber area (Montgomery).
Concentration Ratio	Ratio of radiant energy intensity at the hot spot of a focusing collector to the intensity of unconcentrated direct sunshine at the collector site (Williams).
Control System or Subsystem	The assembly of electric, pneumatic, or hydraulic, sensing, and actuating devices used to control the operating equipment in a system.
Conversion Efficiency	Ratio of thermal energy output to solar energy incident on the collector array.
Cooling Tower	A heat exchanger that transfers waste heat to outside ambient air.

153

Design Point | The specific moment chosen to define for performance characteristics of a system.

Diffuse Radiation | Solar radiation which is scattered by air molecules, dust, or water droplets and incapable of being focused.

Direct Radiation | Solar radiation that comes directly from the sun.

Drain Down | An arrangement of sensors, valves, and actuators to automatically drain the solar collectors and collector piping.

Effective Heat Transfer Coefficient | The heat transfer coefficient, per unit plate area of a collector, which is a measure of the total heat losses per unit area from all sides, top, back, and edges.

Emissivity | The ratio of the radiation emitted by a surface to the radiation emitted by a perfect blackbody radiator at the same temperature (Lapedes).

Emittance | The power radiated per unit area of a radiating surface (Lapedes).

Energy Gain | The thermal energy gained by the collector transfer fluid. The thermal energy output of the collector.

Flow Rate | Velocity at which a fluid travels, usually through an opening or duct (Burt).

Focusing Collector | A concentrating type of collector using parabolic mirrors or optical lenses to focus the energy from a large area onto a small absorbing area.

Fossil Fuel | Petroleum, coal, and natural gas derived fuels.

Gross Area	The total frontal area of a collector, including framing and structural supports (Kreith).
Header	Or: manifold. The pipe running at either end of a solar collector which distributes the heat transfer fluid to, or collects it from the collector (Argue).
Heat Capacity	In a general sense, refers to the ability of material to store heat (Argue).
Heat Engine	A thermodynamic system which undergoes a cyclic process during which a positive amount of work is done by the system; some heat flows into the system and a smaller amount flows out in each cycle (Lapedes).
Heat Exchanger	A device used to transfer energy from one heat transfer fluid to another while maintaining physical segregation of the fluids. Normally used in systems to provide an interface between two different heat transfer fluids.
Heat Gain	An increase in the amount of heat contained in a space, resulting from direct solar radiation and the heat given off by people, lights, equipment, machinery, and other sources (Dresser).
Heat Loss	A decrease in the amount of heat contained in a space, resulting from heat flow through walls, windows, roofs, and other building envelope components (Dresser).
Heat Storage	An insulated container in which energy collected by a solar system can be held for use when the sun is not shining or during extremely cold weather; the heat may be stored in a variety of media including water, rock-beds, paraffin wax, or eutectic salts (Argue).

Heat Transfer Fluid	The fluid circulated through a heat source (solar collector) or heat exchanger that transports the thermal energy by virtue of its temperature.
Heliostat	An electro-optical-mechancial device that orients a mirror so that sunlight is reflected from the mirror in a fixed specific direction, regardless of the sun's position in the sky (Williams).
Honeycomb	Used to suppress free-convection heat transfer across te air gap between a collector plate and its glass cover and to reduce radiation losses from the collector (Kreith).
Hot Spot	The location on a focusing collector at which the concentrated sunlight is focused and the highest temperatures are produced (Williams).
Incident Angle	The angle between the sun's rays and a line normal to the irradiated surface (Kreith).
Incident Solar Energy	The amount of solar energy irradiating a surface taking into acount the angle of incidence. The effective area receiving energy is the product of the area of the surface times the cosine of the angle of incidence.
Irradiance	The amount of solar radiant energy falling on a surface per unit area and per unit of time (Kreith).
Linear Concentrator	A solar concentrator which focuses sunlight along a line (Williams).
Load	That to which energy is supplied, such as internal electrical consumption of each subsystem. The system load is the total solar and auxiliary energy required to satisfy the required functioning.

156

Long Term Storage	The heat storage capacity of a solar thermal system that works on a long term nual cycle.
Manifold	The piping that distributes the transport fluid to and from the individual panels of a collector array.
Mass Flow	The mass of a fluid in motion which crosses a given area in a unit time (Lapedes).
Microclimate	Highly focalized weather features which may differ from long-term regional values due to the interaction of the local surface with the atmosphere.
Nocturnal Radiation	The loss of thermal energy by the solar collector to the night sky.
Operational Collector Efficiency	Ratio of collected solar energy to incident solar energy only during the time the collector fluid is being circulated with the intention of delivering solar-source energy to the system; in other words, when the collector is in track position.
Parallel	Two or more equivalent systems working side by side.
Peak Watt	Unit used for the performance rating of solar electric power systems; a system rated at one peak watt will deliver one watt at the specified working voltage under peak solar irradiation (Kreith).
Process Heat	Heat produced in the form of hot water, steam, etc, for industrial process applications.
Pyranometer	An instrument used for measuring global radiation (McVeigh).
Pyrheliometer	An instrument used to measure the direct irradiance of the sun along a surface perpendicular to the solar beam; diffuse radiation is excluded from the measurement (McVeigh).

157

Radiant Flux Density	The amount of radiant power per unit area that flow across or onto a surface (Lapedes).
Radiation	When unqualified, usually refers to electromagnetic raidation (Lapedes).
Radiometer	An instrument used to measure radiant energy.
Rankine Cycle	A closed heat engine cycle using various components, including a working fluid pumped under pressure to a boiler where heat is added; an expander (turbine) where work is generated; and a condenser used to reject lowgrade heat to the environment (Kreith).
Reflectance	The ratio of radiation reflected from a surface to that incident on the surface.
Reflected Radiation	Solar radiation which strikes an exposed surface after being reflected from the grounds, trees, buildings, snow, etc.
Rejected Energy	Energy intentionally rejected, dissipated, or dumped from the solar system.
Reflecting Surfaces	Usually highly polished metals or metal coatings on suitable substrates.
Reflectivity	The property of reflecting radiation possessed by all materials to varying extents (Kreith).
Reradiation	Radiation resulting from the emision of previously absorbved radiation (Webster).
Scattering	Interaction of radiation with matter where the direction is changed but the total energy and wavelength remain unaltered (McVeigh).
Selective Surface	A surface which has a high absorptivity for incident solar radiation but also has a low emissivity in the infra-red region (McVeigh).

Sensible Heat	Heat stored in a medium (such as water, bricks, or another substance) in which there is a temperature rise (Argue).
Series	Two or more systems connected in an additive manner (Argue).
Short Term Storage	The heat of a solar thermal system that is able to supply heat for a time, short compared to a reference period. (E.g. "short term" for a power plant; a few minutes to an hour, with reference to the day as the typical period.)
Solar Constant	The amount of solar radiation which is received immediately external to the Earth's atmosphere and incident upon a surface normal to the radiation taken at mean Earth-sun distance; it is not a true constant as it varies, mainly due to sunspot activity (Mc Veigh)
Solar Energy	Generally describes those renewable energy sources which directly or indirectly are powered by the sun; they include: direct solar radiation, wind, falling water, biomass, and waves; solar energy is also used to specifically describe solar radiation(Argue).
Solar Engine	An engine which converts thermal energy from the sun into electrical, mechanical, or refrigeration energy(Lapedes)
Solar Multiple	The ratio of the thermal energy, actually collected by the heliostat/collector field of a solar plant, divided by the amount of thermal energy, necessary to achieve the rated output of that plant under design point conditions.
Solar Power Farm	An installation for generating electricity on a large scale using solar energy, consisting of an array of solar collectors, steam or gas turbines, and electrical generators (Crowther).
Solar Radiation	Electromagnetic radiation emitted by the sun (Argue).
Solar Tower	A tall tower, positioned to collect reflected direct solar radiation from an array of heliostats; the top of the tower contains the heat exchange chamber and the hot working fluid is used in a conventional electrical generating system at ground level (Mc Veigh).
Specific Heat	The quantity of heat required to raise a unit mass of homogeneous material one degree in a specified way (Lapedes).

159

Specular Reflection	Mirror-like reflection from a surface (Williams).
Stirling Cycle	A regenerative thermodynamic power cycle using two isothermal and two constant volume phases (Lapedes).
Stirling Engine	An engine in which work is performed by the expansion of a gas at high temperature; heat for expansion is supplied through the wall of the piston cylinder (Lapedes).
Storage Efficiency	Measure of effectiveness of transfer of energy through the storage subsystem taking into account system losses.
Storage Subsystem	The assembly of components used to store solar-source energy for use during periods of low insolation.
Stratification	A phenomenon that causes a distinct thermal gradient in a heat transfer fluid, in contrast to a thermally homogeneous fluid. Results in the layering of the heat transfer fluid, with each layer at a different temperature. In solar energy systems, stratification can occur in liquid storage tanks or rock beds, and may even occur in pipes and ducts. The temperature gradient or layering may occur in a horizontal, vertical, or radial direction.
Sun Tracking	The ability to follow apparent motion of the sun across the sky (Crowther).
Thermal Mass	The heat storage capacity of a structure provided by large quantities of heavy material (Argue).
Tracking Collector	A solar collector that moves to point in the direction of the sun.
Total Radiation	The total of diffuse and beam radiation (Kreith).

160

Transmittance   The ratio of radiant energy transmitted
                               through a transparent surface to energy in-
                               cident on it (Dresser).

Trough Concentrator   Single-curvature (or cylindrical) concen-
                               trator characterized by one plane of symme-
                               try (Kreith).